I0478851

THE INEXTRICABLE INTERACTION
BETWEEN
PRECISION INSPECTION
AND
PRECISION CLEANING
A FIBER OPTIC CONNECTION!

All contents copyright:
All Rights Reserved

Message from the Author

In 1990 I changed career path and returned to the electronics industry. My new employer was, and remains to this day, a true early-pioneer in precision cleaning electronic components. In the early days, a main concern was cleaning difficult contaminants from plastic sensitive printed circuit boards and housings that were not as tolerate of chemicals as they are in these times. Televisions were not as stable as they are now...cooking oil was a contaminant and improper cleaning of an expensive Black and White, and then Color TV could be an expensive endeavor! TV lead to 'transistor radios' which lead to the amazing growth cycle that we know today. Early electronics products failed...often. Being without "I Love Lucy" in 1956 was as 'intolerable' as missing "Walking Dead" is in 2016. Fortunately, electronics is far more reliable. A contaminated fiber optic connection is not!

I was taught there are three major concerns: the first was to understand 'what' was being cleaned and 'what' was the contamination. The second was 'what' was being used to clean. The third was 'what' was the concern or 'what' was lacking with the response to the first and second question! Cleaning an electronic component was a critical task then...as it is now. This analysis, common on the production floor, has not transferred to 'field services': when it does there is an improper assumption that the production floor mimics outside plant. The two are vastly different physical and practical environments, These realities are important aspects carried throughout my work and the concepts in this book.

As my career evolved I encountered many fascinating applications. Overlaid was a global initiative for "Clean Air" and "phase out" of chemical cleaning products that had, literally, industrialized the planet. More so than the environment were concerns about worker safety: limits of exposure. This lead me to understand the concept of 'applications-specific cleaning'. Rather than use 'too much', we developed processes that used less and performed as well or better. These are concerns that continue to this day.

As my career path veered from electronics to fiber optics in 1999: I began to realize the topics of the work you are studying. By and large, cleaning fiber optics has been limited by the ability to actually see the surface: the 1st question!

In July-2014 I retired from a long corporate career. Since that time, I continue research independently in an objective, 'vendor-neutral' stance. This enables free-thought and ideas that are not hindered by a production run or marketing position! This is the 6th book in a series: please consider the information as 'intellectual property' with value. There are many who have contributed to this work by enabling me to share their day-to-day work experiences,

In March-2016 I invented a new way to see the fiber optic surface. The intent is to prove a thesis I have long espoused. Prior to that time _cleaning a fiber optic connection was performed from an inspection point of view...in two-dimensions._

With this work you will study how to inspect from a precision-cleaning-point-of-view...in three dimensions. The two positions are not the same.

All contents copyright:
All Rights Reserved

WHAT'S INSIDE:

Page Number	Contents

Thank you for purchasing this book.

It's valuable we 'baseline' this study with the information below.

These are the existing ways and means that standards, word-of-mouth, and science approach cleaning a fiber optic connection:

1. A DRY PROCESS

2. A 'WET-TO-DRY' PROCESS

3. A "HYBRID OR COMBINATION PROCESS"

4. A "BLIND-CLEANING© Process"

Which works "best"?

INTRODUCTION:

In the 1980's many of us were 'simply amazed' as FAX machines replaced Telex as the preferred means of instant transmission! In rapid succession, consumer, industrial, military, and communications markets introduced technologies that were updated, superseded, and abandoned with astonishing regularity.

About the same time, the first waves of fiber optic deployment began. Barely thirty years later the sciences of fiber optic transmissions still change and evolved. Existing standards for certain phases of fiber optic transmissions remain the same. However, lagging is an understanding of the science of precision cleaning and precision inspection. As of Fall-2016, the last formal standard for inspection and cleaning is Telcordia GR-2023-Core. The most often cited is IEC 61300-3-35 and IRC TR-62627 which are 2008 documents. For any document to be published is, in itself, a 3-5 year task. Clearly, standards for the critical precision cleaning and inspection segment of fiber optic transmissions were obsolete by the time they were published, disseminated and associated equipment produced to support them. Revisions of standards between 1998 and 2008 did not appreciably change our understanding. Telcordia GR-2923-Core, published in 2010. is the latest standard.

As of this writing, there is consideration to relax cleaning standards in the upcoming IEC 61300-3-35 update. This is not a "Best Practice" for field service.

WHY IS IT IMPORTANT?

To understand the answer for fiber optic surfaces, one only needs to realize the importance to society and humanity of cleaning anything! Do we eat from soiled utensils? Is clean clothing important? How often do we rush to mop a spilled coffee or soft drink before the liquid seeps into a cherished electronic device? Do clean eyeglass or vehicle winds screens matter? Did you bathe or shower today?

In most aspects of humanity, clean matters! While a smudge on a window may only be an annoyance, a small dust particle or brush of skin lotion on a fiber optic surface can create reflectance causing signal loss. In the 1980's as fiber optic signals were measured along side of copper in kilobits. Today, while some copper can pass high capacity, there are new 'super channels' of fiber optic transmissions that are passing terabits when only a few years ago gigabits and megabits were considered amazing advances.

While some may claim that a terabit is 'too far into the future', the real issue is training for the future. Existing training programs rarely detail precision cleaning and manufacturers are reluctant to define actual cleaning products or processes beyond the phrase 'clean and inspect the connection'.

In this book you will learn a little about the history of precision cleaning and a great deal about the science. In 2016 a new inspection device was invented that, for the first time, clearly provides images of a fiber optic surface that enables you as installer, network designer, or, trainer to define cleaning processes in an applications specific way. You will learn that existing standards are based on two-dimensional geometry and inspection devices invented in the 16th Century. You will come to better understand that precision cleaning the right way takes no more time than any other way. Your work whether it is for 1 megabit or 100 terabits is performed the same way.

Thank you for purchasing this book. I have spent my entire career, both professionally and in all other ways, working from "worst case". Some of this arises from an life-long SCCA road racing activity where building an engine or suspension 'right the first time' saved time and effort and, assured personal safety and cost less!

Precision cleaning a fiber optic connection likely is not a matter of life and death! However, properly cleaning and inspecting a fiber optic surface assures your work is done properly, your customer is delighted, and assures the future of the medium of fiber optics.

Please join me: In Search of: Best Practice

WHAT DO WE MEAN WHEN WE SAY "BEST PRACTICE"?

No matter where you are now, you are often surrounded by 'buzz words': common phrases used to describe just about anything!

Actually the term "Best Practice" is defined in many world reference books such as Merriam-Webster's and Collier's English Dictionary.

"Best Practice" is defined as:

1.) "...a method or technique that has consistently shown results superior to those achieved with other means and that, in turn, is used as a benchmark..."

2.) "...a best practice can evolve as improvements are discovered..."

3.) "...a means to describe the process *that multiple organizations can use...*"

Existing standards for cleaning and inspection are developed from an inspection point of view! Inspection standards are limited by the technology of both the inspection devices, which although impressive, are simply reiterations of 16th century microscope designs. Cleaning standards tend to consider debris and contamination that can be removed by a given cleaning tool. Participation on standards groups may be a rather expensive effort with fees that limit many participants.

"Best Practice" follows the tenets above: a.) superior results from other means, b.) evolving practices as improvements are discovered, and, c.) processes that can be used by multiple organizations. These tenets also define fiber optic technology.

To these three attributes of a "best practice" I add one more. *"Best Practice is the interaction of the network designer as communicated to the installer as trained by one of many styles.*

Best practice for fiber optics can't be a standard that is updated every 10 years. Even every 5 years is too long a time as our sciences are in continual evolution. It is that reason that I advocate for a RAPIDLY EVOLVING TECHNOLOGY Standard. An RET is a new standard type that is, perhaps, Internet based and annually updated. An RET on a BLOG could be updated in a 'rolling fashion". In this way, inputs from all over the world, in all disciplines can be input to subject matter experts, approved or not, and quickly disseminated for the betterment of all.

In short, "Best Practice" is not a "buzz word", it really exists!

No matter the application, when deploying any fiber optic connection there is a connector with surfaces that must be cleaned and inspected. Future generations will benefit from what we do now.

The 'problem' is two-fold: there are many cleaning products and few connections are actually viewed. The cleaning process itself is ambiguous. Some speak to a "dry process" while others advocate "wet-to-dry". Existing standards are often trained with the instruction that *'if the dry process doesn't work, then use wet-to dry'.* Even with that seemingly fail safe concept there is a further-complicating conundrum: surveys I have conducted over the last 5 years indicate that only 60% of connections are actually viewed prior to and post cleaning. There is still confusion: which test instrument will indicate 'clean'? There is a sense that "since I have a light on the test gear, the connection is clean".

Furthermore, the 'dry process' moves debris and likely does not remove it. A 'dry process' is actually best for a 'fluidic contamination'. 'Wet-to-dry' is a rather obscure instruction as the small and intricate connector assembly can easily be overwhelmed and flooded. At the end of the day, of only 60% of connectors are inspected, how does one know if the surface is actually clean?

Over laid on these concerns are other questions: how much of the connector should be cleaned? Existing cleaning products are based on limited ability of existing inspection devices to see a small aspect of each connector surface. For example, on a 2.5mm end face, which is 2500 microns, existing 400x video inspection can only actually see 250-300 microns of the total horizontal surface area.

This end face is viewed through a RMS-1 TruVideo© microscope using digital photography. In the center of the box is the fiber. The area inside the box is an

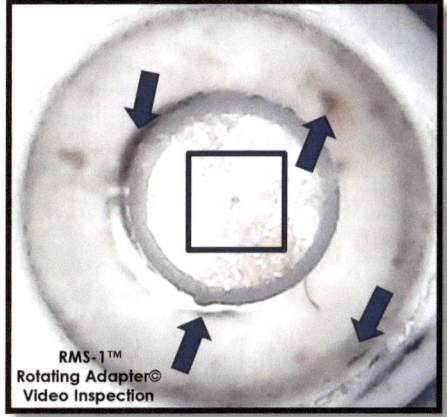

RMS-1™
Rotating Adapter©
Video Inspection

approximation of the image as seen through 400x video inspection. This particular contamination is a fluid, which accounts for the glossy sheen of the picture.

This area is termed the the "field of view". There are three dimensions to this connector, although existing standards only enable view of one: the 'horizontal surface'.

Obviously, there is a third dimension to this connection. The arrows points to the debris 'standing proud' on the 'vertical ferrule'. As you look to the outer edges deeper into the connector, there are darkened areas that are also contamination. Images like this have rarely been seen and they also may be new to you. Obviously, there is more to an end face...than a 'flatland'!

Existing standards define the horizontal area within the box as Zones: 1-2-3. I define the area outside the box on the horizontal as Zone-4. The remainder of the connection is Zone-5. Understanding the science of the connection, the science of inspection, the science of contamination and the sciences of cleaning is the main theme of this book.

You will learn how to properly evaluate images like this and determine that is contamination and what is not. You'll better understand how to inspect and how to properly clean. You will "Future Proof" your work for "Best Practice" not for yesterday or tomorrow...but for each installation in an applications specific way.

PART-1:

THE SCIENCES OF CLEANING

Somewhere in the Middle East, about 5,000 years ago, soap was invented. Some historians credit The Babylonians and others The Egyptians as water, alkaline salts, animals fats and oils were combined as forerunners for what we use every day for many tasks.

This cleaning is known as an 'aqueous process'.

Somewhere in the first millennium The Chinese, along with The Babylonians, discovered the black ooze of mineral oil. Through processes unknown now, the 'ooze' was rendered into oils for lighting...and paving.

By the early 20th Century the black ooze was distilled as a lubricant and engine fuel. Some of the by-products were further rendered into cleaning solvents that created a series of products that created 'solvent cleaning'.

In the 1970's scientists studying Earth's atmosphere noticed 'openings' at both poles and forwarded the thesis of "Ozone Depletion". The ozone layer encircles Earth. It is a deep layer, within the stratosphere, shielding us from much of the harmful ultraviolet radiation from The Sun. The ozone layer became widely known in the late 1980's when it became apparent that certain chemicals, called chlorofluorocarbons (CFCs),worked their way into the stratosphere where, through a complex series of chemical reactions, they destroy some of the ozone. As a result of this discovery, an international treaty was signed in 1978 called the Montreal Protocol. This resulted in elimination of these chemicals and a new series of non-ozone depleting solvent cleaners.

Solvent cleaning is performed for fiber optics using a range of cleaners from IPA to specially formulated cleaners that may be combined with IPA or other chemicals. The advantage of solvent cleaning is the actual cleaner tends to evaporate. The disadvantage is that some of these chemicals may not be plastic safe. Others, as IPA, are flammable. The advantage of aqueous cleaning is these chemicals are plastic safe and non-flammable. Aqueous cleaners require a drying step.

The products used for cleaning anything is a critical selection. The choice of which solvent is as important as which drying material.

The first paper was invented by The Chinese about 100BC...although The Egyptians used the plant papyrus as a precursor. In these times paper wipers are common as napkins and often used on production lines. However, paper is not strong, will tear and leave lint, and must be fortified for precision cleaning tasks such as fiber optics. To do this cellulose (wood pump) is combined with a synthetic material (polyester) into what the industry terms 'non-woven' materials. As well, there are woven materials created of polyester into 'microfibers' that are also used for precision cleaning.

Drying materials are commonly called 'wipers'. While cotton is a highly absorptive material is also can deposit a linting residue as cellulose. Microfibers range from cleanroom to clothing grades. Non-woven combinations of materials are often the best balance between cost and performance.

Each selection of any cleaning material for fiber optics is an 'applications specific choice' in the same way you might select a product to clean your clothing, automobile or other surface.

The Three Dimensions of Cleaning Anything

INCLUDING ALL FIBER OPTIC CONNECTIONS

1.) **What Contamination am I cleaning?**

2.) **Where is the contamination?**

3.) **How do I remove the contamination?**

It's close to an automatic decision: hands are soiled and we select from dozens of brands of soap products. However, if you are an engine mechanic you know the soap you used for personal hygiene isn't very good for the grease on those same hands: so you select one that has a 'kicker'. That 'kicker' may have a citrus aroma and it's called a terpene.

The shirt you wore to dinner has a food stain and the one you wore to work has a grease stain. For one contamination you may 'pre-treat' while the other is a simple task of selecting a cycle on the automatic clothes washer.

Every day, through countless circumstances we select cleaning materials in an applications specific way. This is not the case for precision cleaning fiber optic surfaces.

Inspection and cleaning ... anything ... is a well-established scientific roadmap that began more than 5,000 years ago with the invention of soap and paper. This 'map' is not often followed today when we consider the critical cleaning application that is a fiber optic connection. My research and this book will guide you through 'why and how' to 'best practice' so your work today is as important as it will be to the future.

<u>Hint</u>: Cleaning the GTV Super car and the Fiber Optic End Face is not the same as cleaning a wall, or drain!

Facts & Fictions...

Hey, Bob...do you know how to clean a fiber optic connector?

Hi, Bill! Most times I clean it *under the collar of my shirt* ... that's better than the sleeve ... or dirty jeans!

Really? Someone said just swipe it over your forehead!

The dialogue is a true story. Sadly, frustratingly, there are many myths of precision cleaning that exist not only in our Industry, but also. others as well.

It makes us smile because at one time or another a shirt sleeve was all there was! In other times, fiber optics were not nearly as critical as they are now. In other times a few kilobits or 10 megs was all that was being transmitted. In other times there were fiber optic specialists.

In these times just about every contractor or worker encounters fiber optics...somewhere! In these times megabits have become gigabits which have become multiples and pushing into terabit ranges.

The problem is that "Bob and Bill" are working from an old memorandum! Perhaps they were trained once and the training itself was based on old information or a commercial product rather than 'the science of cleaning'.

Over the years likely tens of thousands of "Bob and Bill" have been undertrained and their knowledge passed on to others. Yes, this is the time for the important and difficult task of re-training many hundreds of thousands of impressions. If we don't 'do this thing', other sciences such as DOCSIS and Ultra Capacity Category Cable can obsolete our science. **Always remember: The Weakest Link in a fiber optic transmission is the condition of the end face.**

If you are an installer, be sure you understand the significance of precision cleaning. If you are a trainer, take time to update your information on a regular basis. The training meeting you attended or developed a year or so ago, may not be relative to the installations of today...or tomorrow. Yes, it is that important!

MORAL OF THE STORY: ELIMINATE CLEANING MYTHOLOGY: STUDY AND LEARN THE FACTS FICTIONS...

1st What Contamination am I cleaning?

The first type is something that is "dry": dust or sand. It can also be something that was once "wet" and has dried into a solid or semi-solid. Something 'dry' can also be a residue from an inadequate cleaning procedure.

Contamination that is "dry" can lay on a surface, or, it can be embedded on a surface which is often called, a 'substrate'. A contamination that is 'dry' can also 'surface bond; to the substrate. This may be accomplished by static field attraction. The dust that, perhaps, is on the screen of the computer may be surface bonded by a static field attraction.

Contamination that is 'dry' tends to stay in place. Generally it will not move unless the surface bond is broken. An example would be dry leaves...an example of 'surface bonding' would be wet leaves!

'Fluidic Contamination" is anything that is not dry! By its nature, fluidic contamination moves about and does not stay in place in the same way as something that is dry. An example would be the drink you spilled and rushed to absorb with a wiping cloth!

The third general type of contamination is a combination of 'wet and dry' types. It may also be 'an unknown' because the percentages of the types is not discernable.

The same common sense we use to select cleaning materials and methods for everyday events are applied to the microscopic surfaces of a fiber optic wisp of glass.

Understanding 'the science of contamination' is important to precision cleaning and inspection. Knowing there are different types is the baseline of our understand. Knowing there are more effective ways to remove contamination is essential. This is performed in a "Best Practice" method.

As fiber optic deployments extend into all aspects of communications in all corners of the world, you as a designer, installer and trainer become aware of the applications specific nature of these installations and deployments. There are infinite contamination possibilities and "best practice" is awareness.

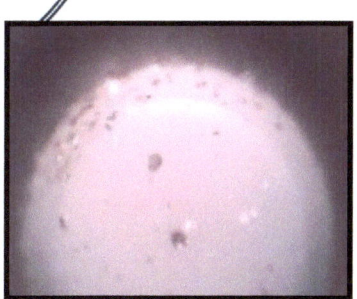

Dry Debris tends to stay in place

Fluids move and can transfer

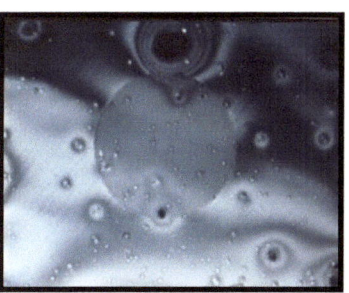

Some contamination is unknown.

2ⁿᵈ Where is the Contamination?

WHAT IS A FIBER OPTIC END FACE?

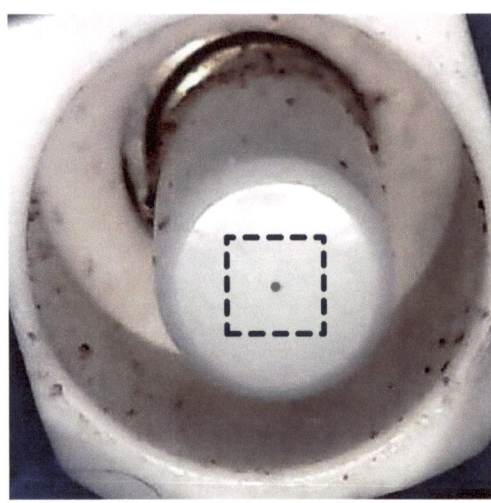

Existing standards define the end face as the area within the dotted lines. You are seeing a digital picture of a 2.5mm UPC fiber optic end face. Obviously there is more to the surface than what is contained within the box! *"Why do we consider it this way?"*, you might wonder. The reason seems to be that the way we have been trained to look at this surface is limited to microscopy in two dimensions, as well as an understanding of contamination in two-dimensional 'diameter'.

Two-dimensions is not 'good enough' to determine if the surface is clean...because there is more to the surface than we are able to see. Until now. Prior to this time, magnification was increased to 400x to see more detail of the contamination. Newly introduced inspection actually backed off magnification to 130x and enhanced resolution to identify contamination. The digital image you see here is an advance using digital photographic imaging which results in a 'virtual 3-D' record.

The total image you see is the 'field of view'. Existing standards concern themselves with the area within the dotted lines and the area within the dotted lines is defined in Zones...usually 1-2-3. The actual fiber is the 'dot' within the square: 9 microns for single mode and 62 microns for multi-mode. While the primary concern is contamination that may cover the fiber, also of concern is contamination on other sectors of the connector that may accumulate contamination that can move to cover the fiber usually in the time of post cleaning and inspection...when the fiber is working! I believe that total cleanliness of the complete connector assures a successful deployment.

I define the complete horizontal end face as "Zone-4". The other area of the 'vertical ferrule' contaminated in this picture, as well as the remainder of the connection, is Zone-5.

If the only contamination was 'dry', then Zone 1-2-3 limits might be acceptable. However, since contamination may move about and transfer, as fiber optics are deployed in many different environments, a higher level of definition of an end face with increased awareness is 'best practice'.

3rd How do I remove the Contamination?

It depends!
ELIMINATE THE
CONFUSION OF CHOICE

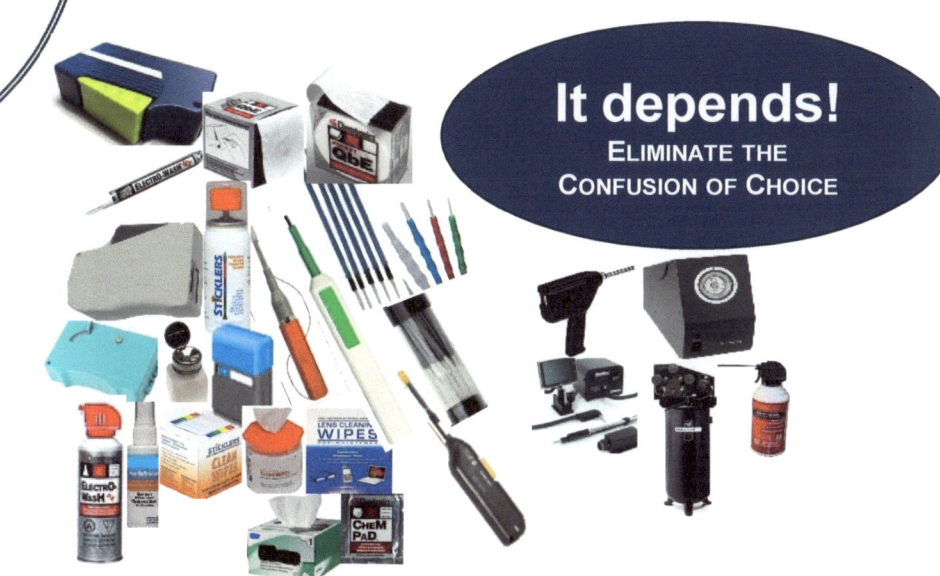

WHY ARE THERE SO MANY CLEANING PRODUCTS AND OPTIONS?

Doesn't one work best? Are they all the same? How do you select which one from an amazing array of products? Should it be the most expensive or will good enough be good enough? Do I use a dry tool? When do I use a solvent? Should I use 99.9% IPA? Do I need a special fiber optic cleaner? What about wipers for optical lenses, do they work? What's a cleaning platform? What's a reel cleaner? What's a probe cleaner? When do I use a 'swab'? What's a stick?

Confusing, isn't it? It seems as though one of these would work better than the other! However, this is not the case as these and many other products all claim the ability to clean a fiber optic connection. The real question is" "what do these products actually clean?". How do they work?

How do you eliminate the confusion of choice and find best practice?

Before making a product choice, I think it's very important to understand what each one does. It's important to understand that some of these products are NOT appropriate for the fiber optic precision cleaning process.

This is the point that you ask the supplier: "what were you thinking?". Remember, your distributor may not be technically able to answer this question. The distributor is the 'supplier' and not the producer of these products. It's a good idea to speak with a factory representative before making a major commitment.

You will also learn that most of these products can easily and effectively 'work better'.

Deployments can be in harsh, diverse environments with myriad debris and contamination types...everywhere. "How to Clean" becomes an applications specific decision based not as much on the 'product' as the actual process.

The first cleaning and inspection standard for a fiber optic connection was published in 1998 as IEC 61300-3-35. At the same time a new cleaning device entered the market and quickly became the *defacto* way to clean. At the time, fiber optics was still new and many connections were in equipment and the need to clean them was on the production line. Most of the equipment of the time was 'precleaned' and since transmission rates were far lower than now, cleaning was more of an afterthought than a necessity. Times have changed; the standards have not!

As fiber optics evolved, so were there advances in copper technology. DOCSIS systems typically used by Cable TV and co-ax and category cable interfaced with fiber optics. CoAx and Category cable connections are not limited by 'the weakest link' in fiber optics: the condition of the end face at the time of test and transmission. If the fiber optic connection is improperly cleaned, then transmission problems occur in the time of post-cleaning and post-inspection. How many times have you heard: "I just cleaned the fiber and it worked'?

The discussion about fiber versus co-ax versus category cable is ongoing and open-ended. For the foreseeable future, all three technologies will continue to evolve as one advances and leap-frogs over the other. However, one aspect of the science will remain the same. This is the technician who is actually working the day-to-day installation. Some of these persons will have greater skills in one segment than an another person. Inevitably, a 'fiber guy' will know things a 'copper-head' does not, and vice versa. We are all in this together.

The most important skill that all can understand is the need to clean and inspect...properly and effectively...each fiber optic connection. It is the one weak link that all can master to the advantage and advance of our mutual success.

What Does It Mean? "Future-Proof"*

➢ The <u>process</u> that seeks to anticipate future developments.

➢ Actions taken to minimize negative consequences. Seize opportunities

➢ Change old perceptions into new realities

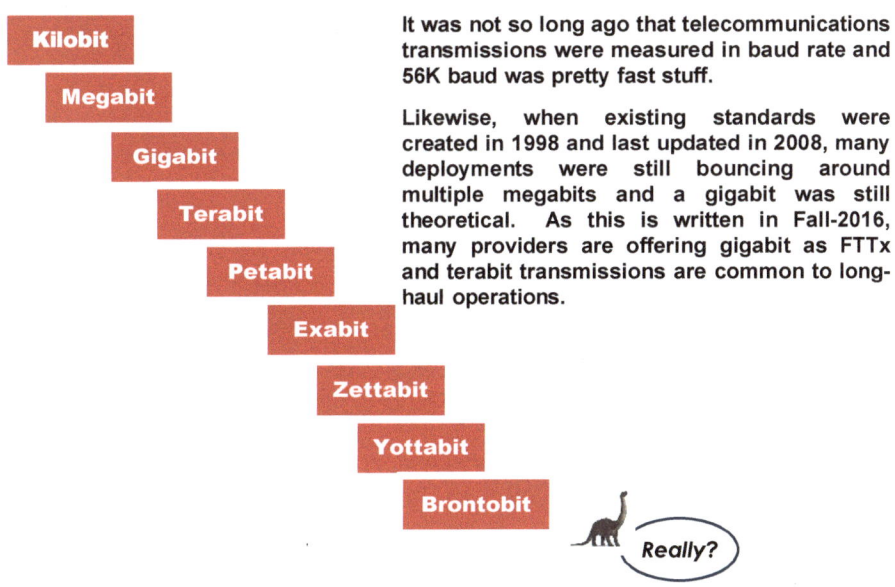

It was not so long ago that telecommunications transmissions were measured in baud rate and 56K baud was pretty fast stuff.

Likewise, when existing standards were created in 1998 and last updated in 2008, many deployments were still bouncing around multiple megabits and a gigabit was still theoretical. As this is written in Fall-2016, many providers are offering gigabit as FTTx and terabit transmissions are common to long-haul operations.

Beginning in 2010, sciences of transmission began to 'step up a notch" as Nippon Telephone and Telegraph in conjunction with The Technical University of Denmark and Fujikura Corporation demonstrated a one petabit transmission over 52 kilometers using an array of 12 fibers.

In July-2014, Technical University of Denmark set a new world record of 243Tb/s over one fiber: in October-2014 Central Florida University and Eindhoven Technical University established a multi-mode record exceeding 200TB/s.

In May-2015 Telastra (Australia) set a new world record for the longest un-regenerated terrestrial fiber optic link covering 10,358km (6,436 miles) between Perth and Melbourne. On October 12th, 2016 Alcatel-Lucent Submarine Networks and Nokia-Bell Labs achieved 65 TB/s over a single mode fiber over 6.600km... "to meet increasing data traffic demand".

While you may not be deploying "super channels" at this time, you will be in the future. Regrettably, you may have been under-trained to design, deploy or train to this level. Perhaps you think that this is not necessary. However, the reality is that precision cleaning and inspection to "Brontobit levels" is something that assures your work today...is 'future-proofed' for tomorrow.

*adjective: British. Oxford® Dictionary. Technipedia®, Cambridge® Dictionary, Wikipedia®

PART-2:

THE PROBLEM WITH CLEANING

WAIT A MINUTE, EVERYONE SAYS WE HAVE TO CLEAN

So far we've spoken about contamination, myths of cleaning, numerous products available to clean the fiber optic connector, and, touched on existing standards.

There are many ways to clean from reputable producers who claim to have the best cleaning product. Standards from IEC, TIA, Telcordia, IEEE and SAE-Aerospace attempt to present how to clean and inspect fiber optic connections of all types. Professional and casual trainers all speak to cleaning on various levels that range from strict adherence to standards to vague instructions that simply say 'clean it'.

There is not one problem with cleaning, but several and they all interact. Existing standards are based on relatively easy to remove contaminants. This is because IEC (International Electrotechnical Commission the world's leading organization for the preparation and publication of International Standards) is largely focused on production line applications and not field services. The production line is a controlled environment. Workers are schooled and processes are repeated oftentimes through many thousands of applications. Contaminants are controlled by filtration and anticipated because of the environment. Existing standards based on IEC 61300-3-35 may not be applicable for field services. Remember, 'awareness is best practice'.

Existing cleaning and inspection devices are based on IEC 61300-3-35. To this extent, there is commercialization that limits awareness of the sciences of cleaning and inspection that are integral to the work you are studying now.

While cleaning is based on easy-to-remove, and existing inspection is limited to a small area of the actual total connector's 'geometry', still there is a better way forward. That way forward is a 'process change". Ultimately, you as a network designer, you as an installer and you as a trainer can implement new standards and should. This is commonly done with Rapidly Evolving Technology with Web Meetings, White Papers and Technical Articles. Organizations such as ETA, FOA, and BICSE offer continuing education credits and certification which are, in effect, re-standardization. This is done to advance the best interests of the fiber optic community.

"The Problem with Cleaning" is that not only are only 60% of connections actually seen before and after they are cleaned, but also, existing inspection does not provide adequate or sufficient imagery to enable the technician to know if the surface is actually clean! Proper cleaning of a fiber optic connection is not difficult, time consuming or expensive. However, no matter what product or technique, if the connector can't be actually seen in it's reality and entirety, there is no possible way to assure the deployment will be adequate or successful.

"The Problem with Cleaning" is that existing standards are minimum requirements that are being promoted as "best practice". This, in no way denies the importance of any standard. However, this knowledge is a 'call to action' for you to assure your work is to a higher standard that protects your reputation and assures the future of the fiber optic industry. Some will remember how 33rpm records obsoleted 78 rpm and then 45 rpm became the new standard only to be replaced by 8-Track tapes and then cassette tapes which gave way to CD's which have been obsoleted by streaming on wireless. Fiber optics is a Rapidly Evolving Technology and existing standards are obsolete if not by the time they are produced and published, then surely in the 5-10 year period between update.

As this is written IEC 61300-3-35 is scheduled to be updated in 2018. It will have important information but some of it will be obsolete.

A Higher Standard proposed to "Future-Proof" your success

The Five Zone Three Dimensional View

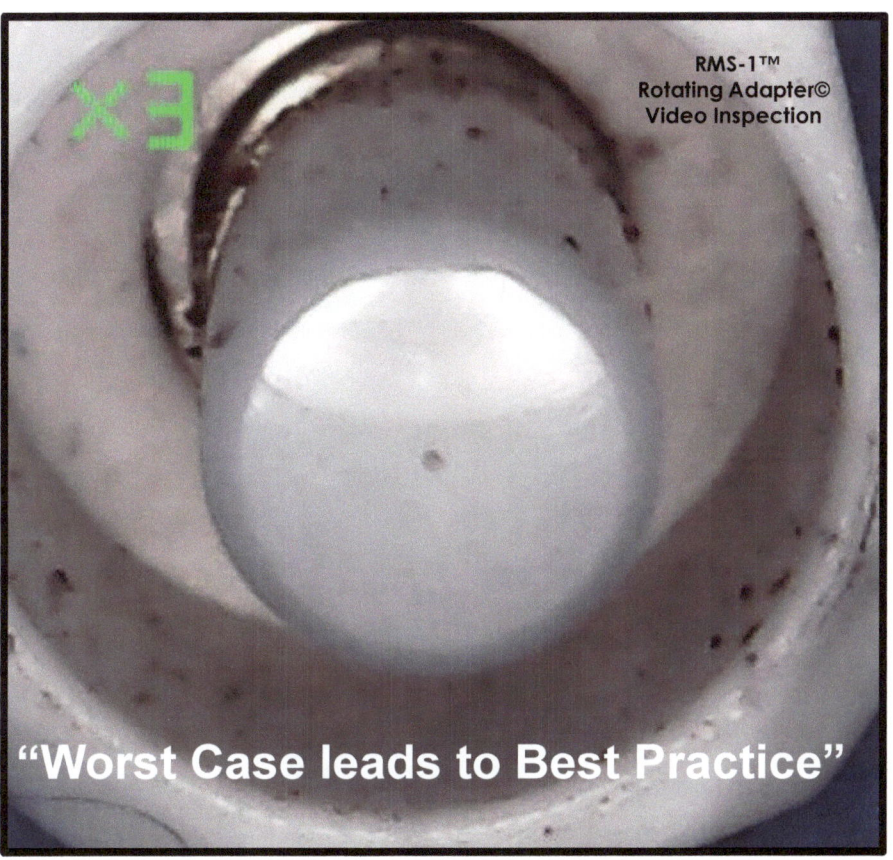

RMS-1™
Rotating Adapter©
Video Inspection

"Worst Case leads to Best Practice"

All contents copyright:
All Rights Reserved

This end face passes IEC 61300-3-35 and will not render a "Fail" by even the most sophisticated "auto detect/pass-fail" instruments.

The core is clear as well as the complete 'horizontal end face surface'. However, there is obvious debris on the 'vertical ferrule' and deep into the recesses of the connector itself.

Can this happen? This is a jumper from my 'test bag' so the debris is real-world. I am also reminded of technicians who were working in Western Kansas and expressed frustration (kindly stated here) about dust from cattle feed lots. Who has not seen images of the dry deserts in places like Afghanistan? These are some of the possible debris...that while you may not remove...you surely benefit by knowing it exists!

This is not possible with existing understanding of contamination and the inability to see it.

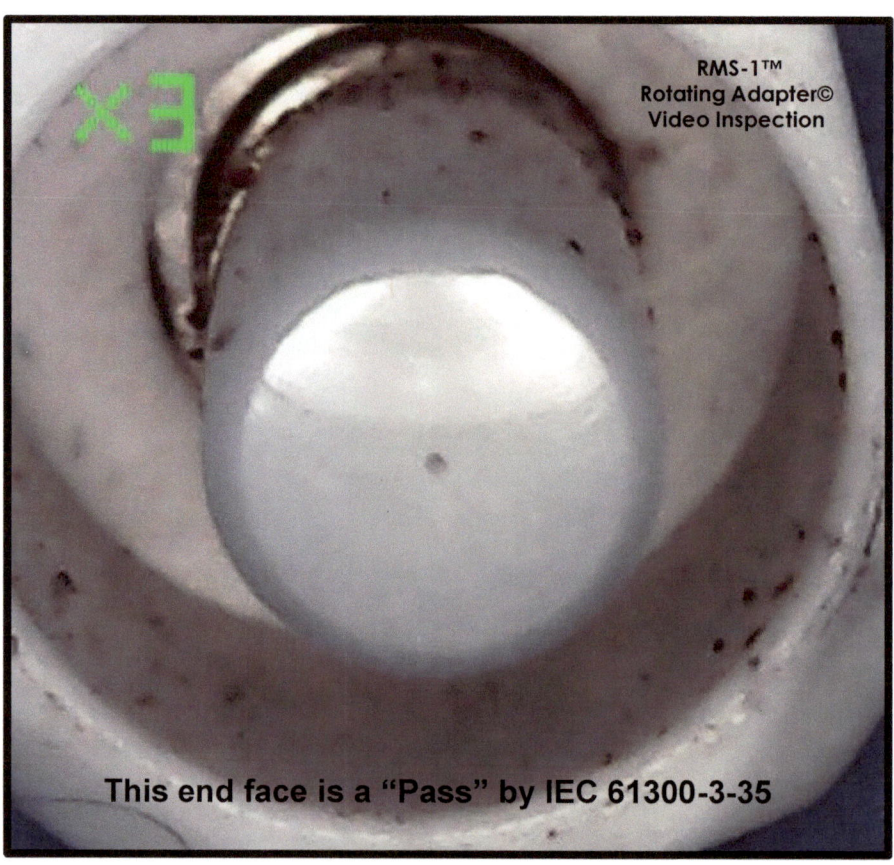

RMS-1™
Rotating Adapter©
Video Inspection

This end face is a "Pass" by IEC 61300-3-35

"Worst Case leads to Best Practice"

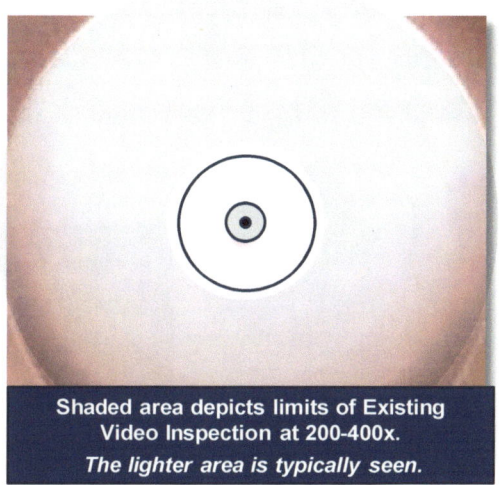

Shaded area depicts limits of Existing
Video Inspection at 200-400x.
The lighter area is typically seen.

I<small>S THE</small> IEC 61300-3-35 S<small>TANDARD FOR</small> V<small>IEWING</small>…<small>AND</small> C<small>LEANING AN END OBSOLETE</small>?

The image above is a clean 2.5mm UPC end face on which are superimposed three rings defining the "core/fiber" as the smallest ring (Zone 1), the next largest ring is the cladding/reflective surface (Zone-2) , and the largest ring is field-of-view typical of a 400x video inspection scope (Zone-3) . At times the 'Zones' are noted as a-b-c-d. The only image seen is a limited surface area of a 'horizontal end face'.

If you use a 'direct view' magnifying device, you may actually see a wider field of view than with 400x. The greater the magnification the more limited is the field of view and the lower the magnification the greater is the field of view.   Magnification increased from 100x in the mid-2000's to 400x.   400x is 'preferred' until about five years ago when a low magnification high-resolution device was introduced.   All of these are, to this day, only showing a two-dimensional view.

Magnification and resolution are important concept: at 100-200x the resolution of the microscope was not effective to determine if the contamination was removeable contamination or an artifact.   An 'artifact' may be an acceptable scratch or other defect in the surface.   Contamination must be removed to 'zero levels'.   The area seen is called the 'field-of-view'. The more you can observe the better the precision inspection and precision cleaning process…and end result.

An advance in video inspection by ODM® a few years ago decreased magnification and increased resolution (sharpness of the image to determine debris.) Likewise a 2016 recent advance in video inspection provides a 'virtual 3-D image'. An interferometer is required to determine 3-D images and is, at this time, cost prohibitive.

Please look at this link on YouTube® to better understand the concepts of 'field-of-view' as well as other discussions regarding precision cleaning,

https://www.youtube.com/channel/UC1a552-2i620UP6mM9WhwRg

Who are they ... and what do they have to do with cleaning and inspecting fiber optics?

Euclid is credited with defining Geometry in two-dimensions about 2,500 years ago. His work was updated in about 100AD by the Greco-Egyptian scholar Ptolemy who observed the stars and realized there was more than a Flatland of two dimensions. In 1492, Christopher Columbus courageously sailed West in the hope of not falling off the edge of a flat ocean. Not only did he discover The New World, but also he confirmed what we understand very well: the world is a 3-D structure.

In 1590 Dutch spectacle makers Zacharias Janssen and his father happened to place two lenses together and invented the first microscope.. The same design, in various forms, is used to this day to measure the two-dimensional nature of everything from an amoeba to a solder joint....as well as a fiber optic end face defined by IEC 61300-3-35.

In the 17th Century, Descartes expanded on Euclidian geometry with his work on spheres. These works were generally considered 'mathematics'. This understanding remained in tact until These are generally considered 'mathematical' science...until the early 1900's when Albert Einstein penned his theories of time and space and Max Planck's study of atomic structure defined them as three-dimensions. In under 2,000 years science evolved from a flatland into what we know today.

Except for fiber optic connections and contamination. IT'S TRUE:

We are inspecting and cleaning a fiber optic connection based on 2nd Century (BC) Euclidean Geometry and Microscopes invented in the 15th Century!

It's time to ... "Go 21st Century" ... with our thinking about these things!!!

> Clockwise: Euclid, Ptolemy, Einstein and Planck, Janssen, Columbus! (Pokémon)

To properly precision clean and inspect…
"Look Outside the Box"
Consider total 'connector geometry'
in Five Zones and 3-D

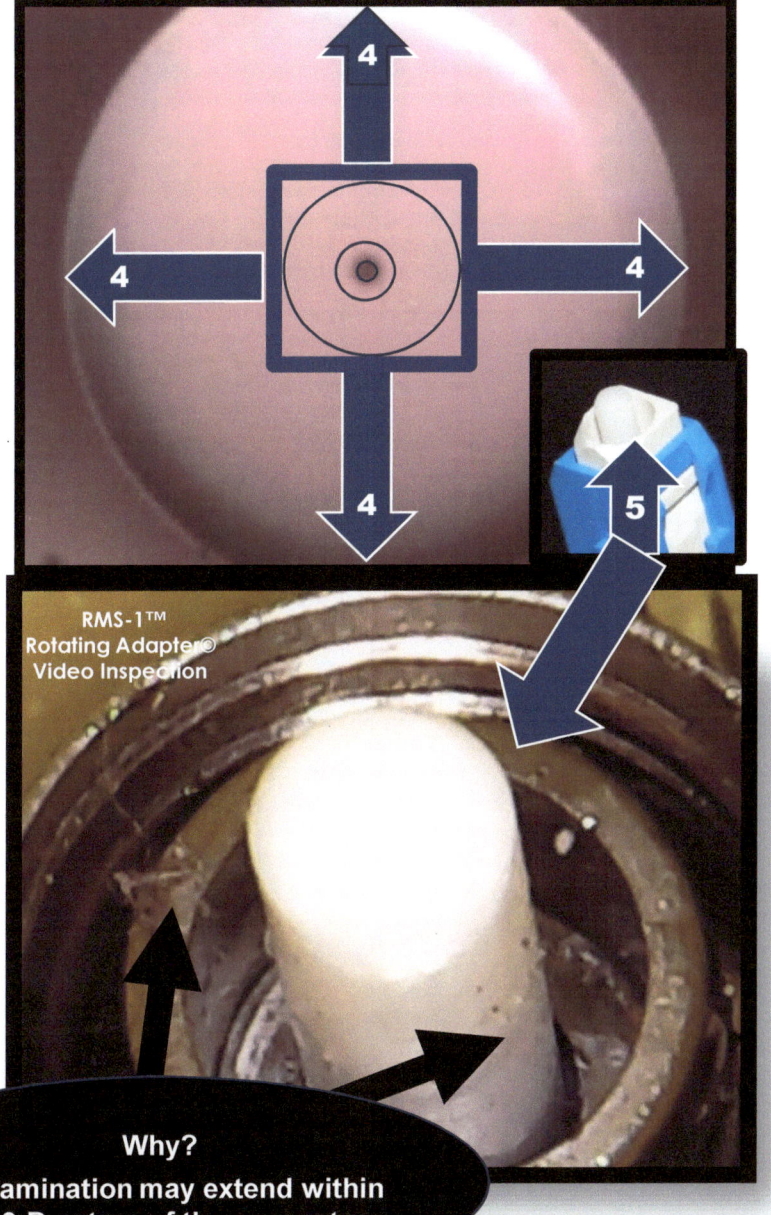

RMS-1™
Rotating Adapter©
Video Inspection

Why?
Contamination may extend within
the 3-D nature of the connector

21

PART-3

HOW TO DETERMINE CLEAN

HOW DO TO KNOW IF A CONNECTION IS 'CLEAN' AND 'GOOD TO GO'?

The Industry has standards that are based on inputs of IEC Committee Members. IEC Committees are formed internationally, meet informally and in formal sessions to input into the standards process. Writing a standard is time-consuming and sober. Since meetings are often held internationally, missing in this process is input from actual users. From the 'mother standard' which is IEC 61300-3-35, TIA (Telephone Industries Association, Telcordia, IEEE and SAE Aerospace write modified standards that more accurately relate to their specific industry.

Once the standard is in effect, various equipment is produced. Since standards are in place for ten years, there is adequate time to develop commercial products that are used to prove the efficacy of the intellectual inputs from some of the brightest minds in the fiber optic industry. Some of the test equipment produced to prove a standard for fiber optic cleaning and inspection include: OTDR, Power Meters, Visual Fault Locators, Direct view inspection and video inspection.

As of this writing, there is a movement to increase the amount of contamination on an end face. While this may be acceptable for a production line operation, personally I believe a 'zero tolerance' is most appreciate for field service operations. I encourage all net work designers, installers and trainers to establish and foster 'zero tolerance ' for contamination on the fiber optic connector. This is done by understanding what is the contamination, where is it, and, may it be removed.

Is this unrealistic? In reality of the science of cleaning, an effective cleaning process can easily, and less expensively be a first time process and not the '5 times' as recommended by existing standards. It may take 5 times to clean using some products or processes. First time cleaning is "best practice" and a proven reality.

CAN STANDARDS BE
OBSOLETE AND TESTING INADEQUATE?

KEY THOUGHT : *"The greater the field-of-view the better the end result."*

How do you know if a connection is clean and 'good-to-go'?

TEST EQUIPMENT: There are two practical ways to know if the to know if the soup stain on your white shirt is there or not: 1.) Someone tells you there is a stain, 2.) you see it for yourself!

In the same observant way, the only practical way to know if a fiber optic surface is to see is to inspect it using either a 'direct view microscope, or, a 'video inspection device'. I categorize inspection into four classes:

    Class-1: Direct View Microscopes
    Class-2: Video inspection
    Class-3: Video inspection with Pass/Fail
    Class-4: Digital Photographic Video Inspection

There are four three other pieces of test gear commonly used by fiber optic technicians: a.) the "Fiber Identifier" indicates which fibers have 'traffic' but will not determine if a fiber end face or connector surface is clean, b.) The "Visual Fault Locator" will determine a break but will not determine of the end face or connector surface is clean, c.) The db Loss Test Set will determine the amount of loss, but will not determine of the end face or connector surface is clean, d.) The OTDR plots the distance to a fault and while the instrument does calculate insertion loss and reflectance which can indicate a clean end face, the OTDR cannot determine cleanliness of other sectors of a fiber optic connection.

All existing Class 1-2-3 video inspection devices are limited to a 'horizontal field of view'. The new Class-4 instrument will be discussed shortly that not only increases the field of view of the entire 'horizontal end face" but also includes the 'vertical end face' as well as other sectors of the connection that include the adapter, alignment sleeve and intersurfaces'

STANDARDS: The Industry has standards that are based on inputs of IEC Committee Members. IEC Committees are formed internationally and meet informally to input into the standards process. Writing a standard is time-consuming and a sober endeavor with input from the brightest minds and influencers in our Industry. Meetings may be on-line, or in various locations all over the world. Missing in this process is input from actual users. From the 'mother standards' which is IEC 61300-3-35, TIA (Telephone Industries Association), Telcordia, IEEE and SAE Aerospace write modified standards that more accurately relate to their specific industry. As of this writing, SCTE (Society of Cable Television Engineers) and BICSI have not published inspection/cleaning standards.

Once the standard is in effect, various equipment is produced. Since standards are in place for ten years, there is adequate time to develop commercial products that are used to prove the efficacy of the intellectual inputs from many hundreds of some of the brightest minds in the fiber optic industry. Test equipment produce products supporting standards

As of this writing, there is a movement to increase the amount of contamination on an end face. While this may be acceptable for a production line operation, my impression is that 'zero tolerance' is most appreciate for field service operations. *I encourage all net work designers, installers and trainers to establish and foster 'zero tolerance' for contamination by understanding there are limitations of existing standards.*

Is this unrealistic? In reality of the science of cleaning, an effective cleaning process can easily, and less expensively be a first time process and not the '5 times' as recommended and permitted by existing standards. Let's take a deeper-dive into standards and inspection.

Is a Standard "Best Practice" or a "Minimum Requirement"?

This is a difficult question and the answer "it depends" isn't very satisfying! Let's go outside the fiber optic industry (for a few hundred words) and try to develop a little clarity! Some of you many be familiar with SAE J-429. *Hmmm, "what's that?".* More of you may be familiar with the grading system used on bolts and nuts: some may have chosen Grade 3 to assemble a lawn mower, maybe used Grade-5 for a riding lawn mower, and imagine you had a racing lawn mower you would have assembled it with Grade-8. SAE J-429 is an 'applications specific' standard and there are hundreds if not thousands of them not only published by SAE (look at a can of engine oil and you will see and SAE-Standard), but other organizations as well such as UL, ANSI, ISO, JEDEC. Google 'standards organizations'...the list is rather endless!

However, within all standards groups is a flaw: the time between publication of the actual standard and the update or re-write. SAE J-429 was first published in 1983 and then updated in 1999 and then again in 2014 to include washers and U-Bolts. This standard, as so many, represents a stable process.

I propose there is another type of standard: this for Rapidly Evolving Technologies. These "RET" include solar panels, windmill farms, electrical vehicles, micro-devices, and fiber optics...among many others it's sure. The "RET" can't wait ten years for an update: technology moves too quickly. The last time IEC 61300-3-35 was updated was a 2008 publication which means the actual work of inputs was finished in 2007 and based on 'those times and technology'. There is another problem: the commercial side of any and all standards as producers than creates a product to 'meet the standard': often an immense investment.

Standards are not often created with input from the most valuable resource: those in the field who have the actual day-to-day experiences unfiltered by third party interpretation. What's the answer? One could be a "RET" standard based on the Internet with global input solicited using existing standards bodies as the 'final authority". In this way, the standards meetings are vetting actual real time input and in these ways a "RET" could be updated annually and always in synchronization with evolving technology.

For fiber optics, which regularly passes from 'theoretical to practical', the concept of an RET can be an important advance.

Eliminate existing standards: ABSOLUTELY NOT! This is the time to think in different ways to assure the future of our science. In the mean time, we continue training and thought development by publishing books as this, writing white papers, and attending Internet seminars. Include clear instructions in your work orders that anticipate 'problems' and be sure you are trained. Until they are updated, consider all existing standards as 'minimum requirements' and not necessarily "best practice".

Don't be stale as the sciences of fiber optics is ever-changing and vibrant. A fiber optic deployment is not the same as a Grade-8 bolt!

2nd **Where is the Contamination?**

A NEW VIEW: THERE IS MORE THAN AN 'END FACE'. Each connector type has its own 'geometry' with various sectors and segments. Any point can house contamination.

This image of a 2.5mm connector end face shows the 'clear-core' along with a clean 'horizontal end face'.

The 'vertical ferrule' is soiled as are the recesses of the ferrule: this debris can transfer and migrate.

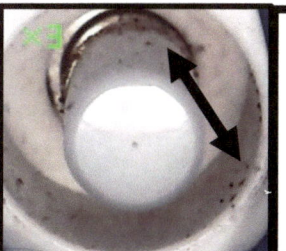

Cleaning each fiber on 'MT-Type' is critical. *Dry debris is present in alignment pins and 'inter-surfaces'.*

Who among us has not 'missed' when attempting to insert the jumper into the alignment sleeve!

Contamination on the adapter can be passed to the backplane connector!

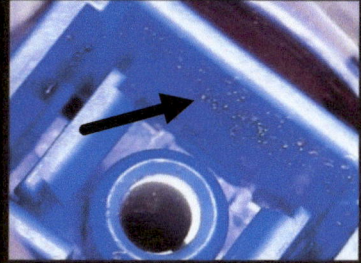

The sciences of fiber optics continue to evolve. This is a 'military style' expanded beam with a 'ball lens' over the fibers.

Demonstrated are contaminated inter-surfaces and alignment pins.

The 2-D View of a connector and contamination (AS PRESENTED IN ALL EXISTING STANDARDS) is not "Best Practice"

Position and Transfer of Dust, Fluid and Other Debris

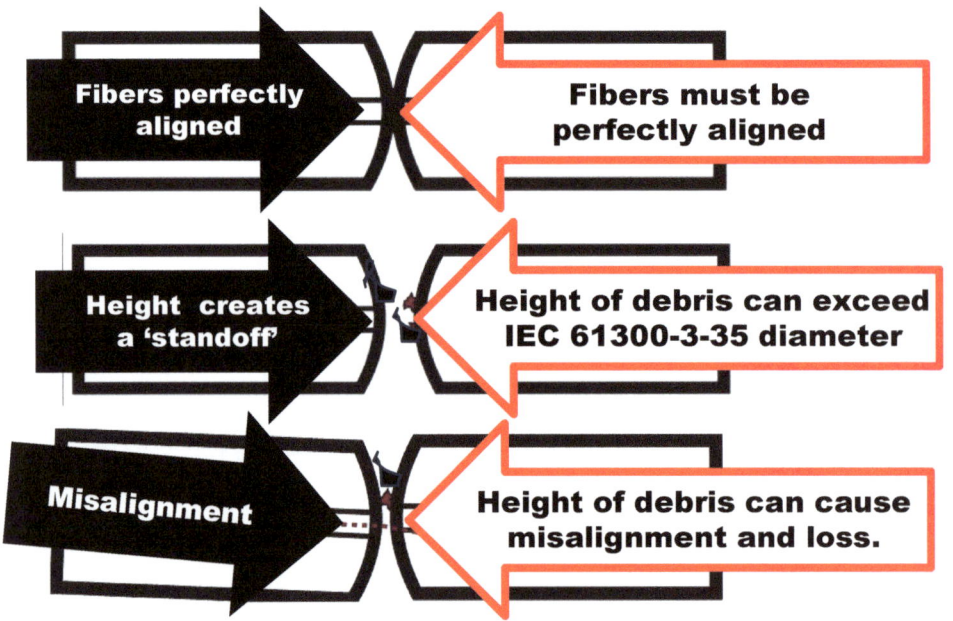

Transfer on contamination and the effect of debris on alignment are recognized with concern from all of us!! The images you see are captured with a new video inspection instrument. The machine was developed in Spring-2016. After years and many miles of speaking about this, I realized there was not a 'proof of concept'. While we clearly understand we exist in 3-D, for fiber optics, there was no practical way to prove it.

RMS-1 is digital photography of the end face based on a patent-pending concept of a rotating adapter™. The rotating adapters enable both the standard IEC 61300-3-35 view but also a totally new perspective. The color pictures are not the same as the back-lighted monochrome images from existing video inspection.

WHICH FIBER SCOPE WORKS BEST?
AMAZING FEATURES FOR THE FUTURE
...WHICH IS NOW!

Whichever instrument you choose, visual inspection is the only way to assure the connection is actually clean.

Which one works for which application? Let's break video inspection into four categories:

Class-1 are direct view microscopes typically with 100-400x magnification

Class-2 are video inspection with or without recording typically with 100-400x magnification. Typically, these are monochromatic instruments with back lighted images and limited field of view of the horizontal surface.

Class-3 are video inspection with automatic detection of contamination and location in a 'pass-fail' mode. Typically, these are monochromatic instruments with back lighted images and limited field of view of the horizontal surface.

Class-4 Video Inspection is the newest design utilizing digital photography and the ability to determine a virtual 3-D image of the horizontal surfaces, vertical surfaces and other aspects of the connector.

Class-1 Direct View Microscopes should not be used to observe any fiber optic connection that has any possibility of an active transmission. Permanent eye damage is possible even with those instruments that claim to have 'safety filters'. These are ideal for cable assembly and checking jumpers as in QC.

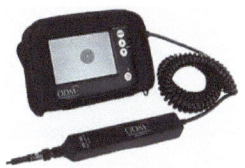

Class-2 Video Inspection instruments are available as in designs ranging from 'entry-level' to highly sophisticated Select from lower magnification, wide field-of-view, or and higher magnification models. Some have still image recording and automatic defect measurement. These instruments only see a small area of the horizontal end face.

Class-3 models may be 'wireless' IOS or Android with automatic defect detection based on IEC 61300-3-35. Most have still digital recording. These instruments only see and analyze a small area of the horizontal end face using algorithms to determine the extent of contamination.

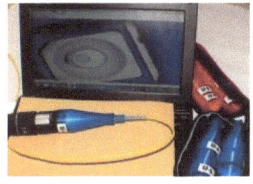

Class-4 instruments feature digital photo imaging and the ability to see the IEC 61300-3-35 standard view plus the remainder of the horizontal end face, the vertical ferrule as well as sectors of the adapter never before seen that can be contamination points. The instrument captures both still and motion digital images in six magnification levels.

THESE ARE NOT "NEW IDEAS"! Early in the research of cleaning a fiber optic connection we encountered Cisco Systems and this internal standard. What is unusual is Cisco looked 'outside the box' and sought to clean contaminants that were not typical. At the time, IEC 61300-3-35 was in the 8th year of publication. This cleaning standard specified "Arizona Road Dust" and "Vegetable Oil" as the base.

There are several grades of Arizona Test Dust. One of the more troubling aspects of this debris, although measured in size, is the cost which is approximate $575 for 20 grams. Clearly, vegetable oil is more affordable.

Cisco took up on the idea of 'affordable' by adding debris noted in the graphic below. What Cisco sought was first time cleaning of the complete 'horizontal end face'. At the time, the only video inspection capable of seeing more than the customary 300-400 micron radius of the fiber core was a MicroEnterprises® ME-2500 with a 'scroller' which would move the camera over the complete horizontal. I am fortunate to own and use a MicroEnterprises ME-5000 which has scrolling capability. I use this instrument to check development of RMS-1 and RMS-2 video scopes.

IN 2006 CISCO® ISSUED AN INTERNAL STANDARD FOR CLEANING A FIBER OPTIC CONNECTION.

An important document using widely varied debris and contamination from the 1998 and, then pending, 2008 IEC 61300-3-35 standard.

Contaminants included:

1.) Vegetable Oil
2.) Metal Shards
3.) Powdered Graphite
4.) "Duffy Dryer Lint"
5.) Simethicone
6.) Arizona Road Dust
7.) Air Dried water
8.) Air dried IPA
9.) Air dried salt water
10.) Finger Oil

Each contaminant was tested ten times using any specific cleaning tool or process.

Successful cleaning was one that removed the debris the first time over the complete horizontal end face

The actual testing is tedious. There are specific methods and procedures for contaminating the end face so that each of the ten efforts was balanced and unbiased. This test set the foundation for product development and my personal understanding of cleaning these surfaces.

What Cisco also established was that there are many possible contamination types. Some of the easiest to remove were Arizona Test Dust and vegetable oil. Some of the more difficult were dried salt water and dried IPA. *Cisco also set the standard for first-time cleaning.* The Cisco Series, nor IEC, or, any other body tests using combinations-of-contamination.

For me, one of the most valuable tenets was the standard to clean ten times. I encourage those interested to use debris that you encounter in this process: if the cleaning product or procedure successfully will clean ten times, then it is "best practice". I am troubled by the cost of Arizona Test Dust and suggest "silica flour" as an alternative. Although, the dust, any dust you are encountering is acceptable,

A copy of "first time cleaning tests" against these contamination types is available from ITW Chemtronics

Why is First Time Cleaning Important?

Of course, a logical answer is "it's a time saver", which it is! However, thinking a little deeper, if the product cleans a wide range of contamination 'the first time' then that becomes an indication of cleaning ability. That was what Cisco was seeking: a Best Practice technique. So do I.

In December, 2014 I conducted a series of 'vendor neutral' comparisons using the same esoteric Arizona Road Dust as IEC specifies plus a fine dust from Afghanistan. Since that time I have added 'silica flour' to the test regimen as this fine powdery dust emulates desert dust. I urge you select dry debris you encounter and follow the 10x testing to determine an 'applications specific best practice' for your specific environment. In addition to testing with 'dry debris' and 'fluidic contamination' the two basic types were combined into a third 'combination type'. This was done because it's likely contamination you encounter might be dust from a data center and hand oil from French fries at lunch! Always include combinations is your testing.

My goal was to create "worst case". This also follows tenets of, for example, SAE J-429 where a Grade-8 bolt, nut, washer and lock washer form the highest quality connection. "Worst case leads to Best Practice": an important tenet as fiber optic transmissions continually increase to match customer expectations and demands.

Debris and Contamination included:

1.) Vegetable Oil

2.) 10-30w SAE Engine Oil

3.) Arizona Road Dust

4.) Desert Dust from Afghanistan

5.) Jergens® hand lotion

6.) Vegetable Oil and Arizona Road Dust

7.) 10W-30 Engine Oil and Desert Dust

8.) Hand Lotion and Arizona Road Dust

- IBC® Tool
- ClePen Tool™
- FerruleMate®
- FerruleMate-2
- Sticklers® 2.5mm Swabs tools
- ITW Chemtronics® 2.5mm Swab Tools
- QbE® Cleaning Platforms
- Stickers® Clean Wipe Platform
- HandiMate™ Platform
- Stickers® FiberOptic Fluid
- Chemtronics® "MX™ Fiber Optic Pen
- Chemtronics PX™" Fiber Optic Fluid

Each tool and product was used per manufacturer's instruction. Tools that were used 'dry' performed better on fluids than on dry contamination. This lead to the conclusion that 'dry cleaning' is a 'mopping action' and not appropriate for all contamination types. Over 90% of the tools worked better using a technique not included in manufacturer's instructions.

"A Comparison Study of Precision Cleaning Methods for All Fiber Optic Connections".
www.amazon.com

Before we discuss precision cleaning, let's return to the discussion about "precision inspection' and specifically the Type-4 inspection device.

It is true that existing inspection for fiber optics is based on 16th Century microscopy. Over the last ten or fifteen years, reputable producers have increased magnification from 100x to 400x and greater. The increase in magnification is to enable discernment of contamination. This is important since there are defects such as scratches or blemishes in a fiber optic end face that can be confused with contamination. As well, an improper or inadequate cleaning process can result in a 'removable residue' being mis-characterized as an 'unremovable artifact'..

With precision cleaning sciences as a 'background" now we turn to 'precision inspection'. Existing inspection is always better than no inspection at all. In fact, one of the most significant and understated tenets of IEC 61300-3-35 and all standards based on
"100% inspection of every connection every time it is installed,
serviced or otherwise opened".

While existing video inspection is better than none-at-all, these instruments are limited to a two-dimensional image of a limited surface area. The innovation of RMS-1 and RMS-2 is the versatility to see both the heritage IEC standard as well as new views that present a virtual 3-D picture of all sectors of the connector. Some of these pictures of both 'direct contact' and 'military style expanded beam' better define the innovation and express the need to clean more than the 'horizontal end face'. Before invention of this video scope these images were not possible to be seen. As you view the pictures, observe the complete connection and presence of contamination which can be removed.

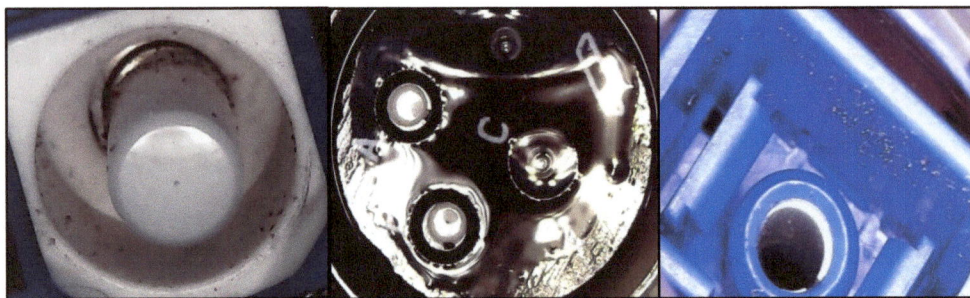

These images show areas of the connector that have rarely been seen. On the left is a clean 2.5mm horizontal end face with contamination on the vertical ferrule and throughout the connector. In the center is jet fuel on a 38999 military style connector: the jet fuel (fluid) will transfer more readily than dust. To the right is a soiled adapter and looking closely there is dusty contamination on the alignment sleeve that can be pushed to the back plane side.

There are more than 600 pictures as these, including unique motion video capture that serve as 'proof-of-concept'.

Seeing as much of the connector as possible is an important advance. Knowing where are 'contamination points' saves time and effort. It's not that each point must be cleaned, it's understanding where the problem is before you can fix it!

These are the existing cleaning processes.

How to determine which one works best, in which application, and understanding a "best practice" are the remaining topics in this book.

1. A DRY PROCESS ... that may only "move" debris and/or a static field

2. A 'WET-TO-DRY' PROCESS ... which may/may not actually dry the surface

3. A "HYBRID OR COMBINATION PROCESS" defines amount of fiber optic grade solvent and a drying technique.

4. A "BLIND-CLEANING© Process" leap-of-faith ... without benefit of adequate or any video inspection

The Industry generally speaks of the first two means. IEC 61300-3-35, TIA 455-240, IEC TR-62627 list these as the preferred ways to clean. These are the oldest standards in publication as of this writing-Fall-2016. IEC 61300-3-35 in in 'update' and will re-publish shortly. However, their efforts are to increase the amount of contamination at a time when 'zero-tolerance' is 'best practice'.

The "Hybrid/Combination" process is published and preferred in the most recent standard: Telcordia GR-2923-Core, published in 2010.

Regrettably, the 4th means is all to common. Perhaps as great as 60% of connectors in the field are 'blind cleaned'.

PART-4

How We Do and Should Not...
... Should and May Not ...
Clean and Inspect
a Fiber Optic Connection!

3rd: What process, tool, or device will successfully remove (any and all) contamination?

I believe and propose to you that precision cleaning any fiber optic connection is an 'applications-specific' task. Success is based on understanding of the contamination, location as well as proper selection of a cleaning product. However, successful cleaning is not the result of a product, *but rather how it is used*.

Successful cleaning is a process...based on product selection. Just as removal of a Philips® screw with a 'straight-blade' driver is unlikely, so is successful cleaning with a product never intended for the task! Yes, some of these exist and are commonly used.

Situational Awareness: Know before you do to the work site:

For those in field services or 'outside plant', each work site is different. Each day is a different ambient and conditions vary by season and region. Sure, this is obvious but as well, these factors influence successful precision cleaning...and inspection. After all, your job to precision clean a fiber optic end face is not the same as cleaning a grease stain from the floor or coffee spill. Sure, this is obvious: it's mentioned because many of the cleaning processes we perform on a day-to-day basis are the result of not only personal 'trial and error', but also benefit from more than 5,000 years of product development! Regrettably, some of the 'science of cleaning' has not been passed to fiber optic precision cleaning.

Many continue to clean as though this is 1998...and that is not 'best practice'.

Before leaving the garage, always know or anticipate: what kind of connector will I work with today? How many are there? What do I have to clean them? Have I been updated on how to clean? Where are they located and how accessible are they? What contaminants can I anticipate? Do I have a video scope to see what I am cleaning?

Who Specifies the Precision Cleaning Process?

Today is your first day on the job and you have been 'promoted' from category cable installer to fiber optics. You are enthusiastic and a little concerned. You attended one CWA® session and have a handful of BICSI® CECs and are working on FOA® and ETA® certification. There is a lot of work. "Isn't there a better way?", you think as you drive to the work site.

You ask yourself: *"I wonder if the network designer as specified a cleaning process for this job?".* The answer is 'likely not'.

But..."why not?". I believe that a successful installation begins with the network design and suggestions for precision cleaning should be an integral part of that blueprint. Network or system designers have immense talent and this work is protected by including a simple recommendation on how to clean and inspect.

Courtesy of:

Do you ever wonder?

"Why are there so many different cleaning products?"

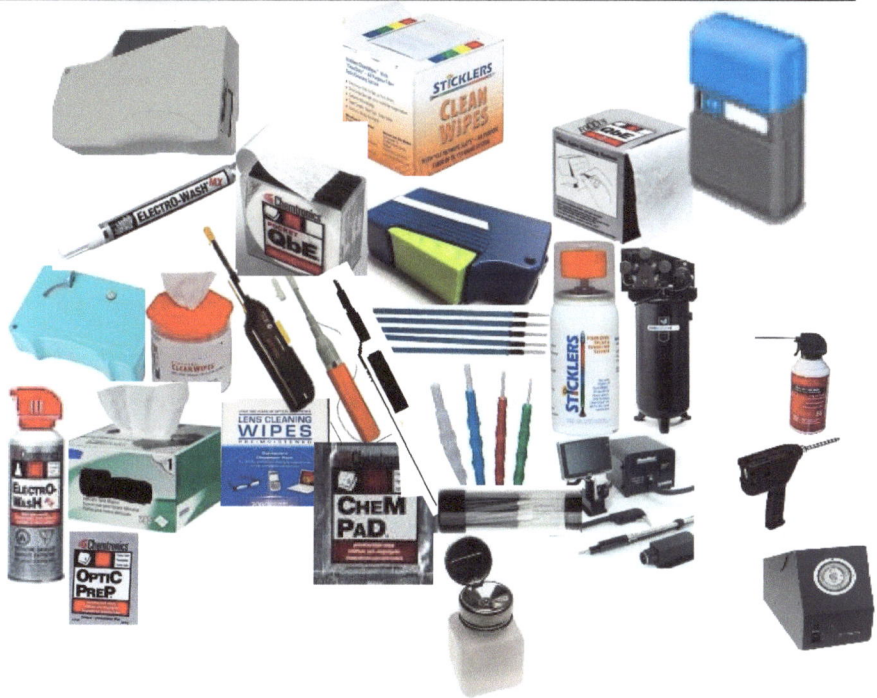

Did you every consider?
"Maybe cleaning is <u>not</u> about product choice
It's about:
<u>How the product is used."</u>

The Confusion of Choice:

There seems to be no limit on product selection! Each one of these tools or devices claims to be the best. What if they are? What if they are not! There are some of these that are not suitable for cleaning a fiber optic end face. How do you know which?

The only way is to understand not what they are...but how they can work better. Let's discover which cleaning process works best. From that you adapt the product in an applications specific way to achieve "Best Practice".

Let's talk about...how to select a cleaning product

RULE NUMBER 1:

Understand the limitations...

Graffiti removal is a different cleaning process than "doing the laundry' and polishing this super car is a different process than polishing the floor!   Each of these surfaces may be contaminated with a different debris!   Stains are pre-treated on laundry in a different way that grease is removed from a floor or painted graffiti from a wall or tree sap from the surface of the supercar!

The current approach to precision cleaning a fiber optic connection is *'if one way doesn't work, try another'.*   However, the first cleaning process may damage the surface or set the debris so that removal is difficult or impossible. "We" know these things...but do not apply them to our science of precision cleaning fiber optics. Until now.

The fact is that existing standards such as IEC 61300-3-35 are based on the production line where contamination is the same, day-after-day, workers are trained and skills honed by the repeatable actions of cleaning the same thing, perhaps, many thousands of times,   The production line is not "our world" and this awareness does not invalidate existing standards. Knowing the limitation advances all to a higher level.

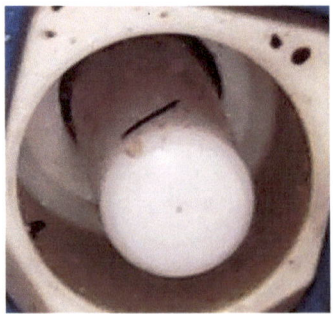

How this surface is cleaned is an applications-specific decision.   To best understand this, let's review the existing cleaning procedures as you may have been trained.

Perhaps you will agree that "Best Practice" isn't more difficult, more expensive, it's just the right thing to do. Starting now.

**Dirt & Neglect
#1 Cause of
System Failure**

PRODUCT SELECTION

What works...
What not so much...
and why?

LEADS TO BEST PROCESS

Is there something "wrong" with IPA?

Likely you have heard the word: hygroscopic. Perhaps you were told *Isopropyl alcohol is hygroscopic.* The word is a scientific term that means IPA attracts moisture to itself. IPA is one of the few chemicals that has this property. *A little known but significant <u>disadvantage of 99.9 IPA</u> is that it attracts moisture to itself at a very rapid rate: it is self-destructive.*

Storage of 99.9% IPA contributes to degradation which begins within 15 minutes of an open container being exposed to the atmosphere. "We always close the top", however the 'headroom (1.) in the container, which increases each time the container is opened, exacerbates and hastens the downgrade from 99.9% to 70% 'drug store" grade. However, even in the purest form, 99.9% 'reagent grade", IPA is still limited in its cleaning ability. Of course, IPA has an excellent sterilization factor and IPA can be used to increase performance of some chemicals. However, 'sterilization' is not the same as 'cleaning'. Never use IPA that has been left in the top of a Menda® bottle and always empty the bottle every day. (2)

In these times of 'environmental responsibility', IPA is added to non-ozone depleting cleaners which do not have effective cleaning ability on their own. Some chemical companies formulate the 3M® invention HFE-7100/7200 with IPA to increase the cleaning ability of both components.

Even so, IPA is limited to clean a narrow range of contamination. For example, IPA was used in electronics to clean rosin flux. IPA would be good to clean a salt residue or hand oil. IPA may not be effective to clean a hand lubricating lotion, mineral oil, fuels

As IPA is 'squeezed' or 'pumped out' of a refillable container, air is brought in. The 'void' in the container is outside air that diminishes the cleaning ability by replacing what as 'reagent grade' with outside air.

IPA is acceptable for fusion splice prep and some pulling gels will successfully clean with IPA. However, "Best Practice" is to reduce the chemical inventory to as few choices as possible. There are several highly effective fiber optic cleaners that perform better than any grade IPA and cost at about the same level. Always challenge the manufacturer...and remember...your distributor is likely not the actual producer of the product. The distributor is a sales agent. Ask for the manufacture's rep...call the factory!

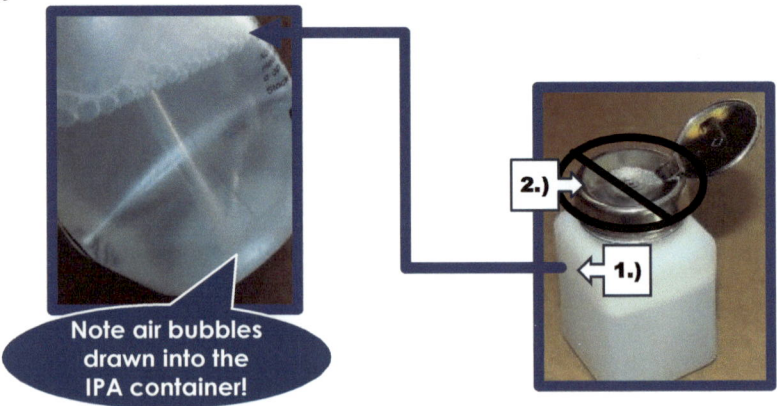

Note air bubbles drawn into the IPA container!

A CLEANING SURFACE SUCH AS A PROBE OR SWAB TOOL HAS SEVERAL CHALLENGES:

1.) Is the surface large enough to actually clean the complete surface?
The arrows point to the width of the cleaning tape. Things to consider:
- Is the tape or swab tool wide enough to clean the end face?
- Is the tape or swab tool wide enough to clean the alignment sleeve?

2.) Is the tool tip sufficiently compliant to compensate for an APC surface?
- Can the tape surface clean beyond the contact zone?

This image shows the bevel on a UPC and points to some of the challenges of precision cleaning. The darkened area may not be cleaned.

THE SMALLER THE CLEANING SURFACE:

1.) The less surface area it will clean

2.) The more difficult it is to absorb and remove fluids

3.) The more times you will use the tool to clean

4.) The less likelihood there is for first time cleaning: Best Practice.

Knowing this is not a reason NOT to use the tool.

Knowing this is information HOW to use the tool.

Precision cleaning products are not "one size fits all".

PRECISION CLEANING IS AN APPLICATIONS SPECIFIC TASK.

Product Limitations: Wiping Materials

<u>WHY NOT COTTON? WHY NOT PAPER? WHY NOT POLYESTER? WHAT IS MICROFIBER?</u>

Cotton is nearly an 'ideal' material to absorb a fluid. However, by nature, cotton has a linting surface that is unpredictable and therefore not acceptable for precision cleaning. Cotton would not be used in a Class-1 Operational Cleanroon for solid state devices and unlikely for medical device production. I believe these are the same standards that should be used for fiber optic precision cleaning,   Think about it...if cotton isn't used in a cleanroom...it should not be used in a data center or top of a bucket in a storm in Oklahoma or 'gentrified' townhome in 'the city"!

<u>WHY NOT PAPER?</u>

Paper is absorbent, but it does not have tensile strength. (While paper wipers are used in come electronics applications, those surfaces do not transmit light and require zero tolerance to contamination as does a fiber optic end face.

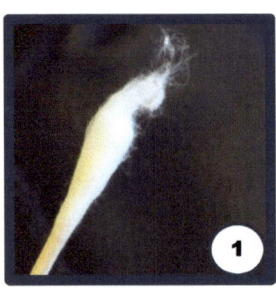
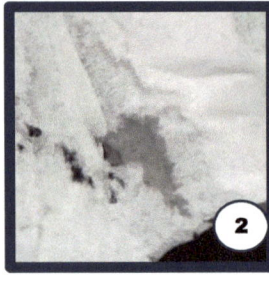
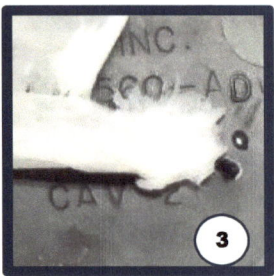

Note the cotton tip in Figure-1. This is a 'low-lint' version and the wisps of cotton fiber could cause havoc on a fiber optic connection!   Note the picture below: this is a photo of a cotton thread 'captured by luck' during a test session

Figure-2 is an example of the 'tensile strength' of a paper wiper...maybe you have a green box of them on your work bench. These are intended to "mop" spills...not for precision cleaning. When the paper wiper is moistened, it becomes a wad of paper. (Firgue-3)

<u>WHY NOT POLYESTER?</u>

Nearly an exact opposite of paper is polyester: it is exceptionally strong, However, 100% polyester is not absorbent. There are better choices.

<u>WHAT ARE MICROFIBER?</u> There are many microfiber types ranging from clothing grades to upholstery grades to cleanroom grades for fiber optics! Microfibers are combinations of polyesters, nylon, Kevlar, Nomex an other materials. Select the right stuff.

<u>WHAT ARE "NON-WOVEN MATERIALS"</u>   The garment you are wearing was 'woven' on a loom of some type!   Non-woven materials are a class of wiper that are created in much the same way as paper: a slurry that is flattened and dried into a sheet. This process is called 'hydroentangled'.

As is the case with micro fibers, there are many non-wovens. For fiber optics, I was once told there are more than 4,000 combinations of polyester and paper from which to select. Some have more paper and others have more polyester: compare, you can feel it. There are several non-woven fiber optic cleaning tools and wipers on the market...right now! They work very well.

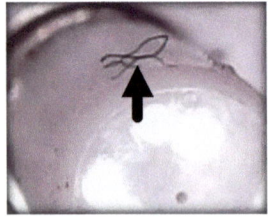

Understanding what you are choosing is a critical task when it comes to fiber optic precision cleaning. Even the grocery story may ask: paper or plastic!

When someone wants to sell you a 'wiper', now you are informed: paper, polyester, microfiber or non-woven? These materials are not the same!

Product Limitations: Fiber Optic Solvent Cleaners

Is a solvent necessary to successfully clean?

As with so many questions, the answer is: "it depends"! Some may disagree, but I take comfort in understanding "precision cleaning" based on not only my personal experiences, but many thousands as you have encountered. Ask yourself: *"Does a dry wiper do better cleaning dust from furniture, or, is the job better with a little furniture polish/cleaner?"*. What about trying to wipe a coffee stain from your shirt? *Isn't it more effective to use a little water and maybe soap?* On the other side of the question: if you spilled coffee, would you use a wet cloth to 'dry' it? These are 'fundamental tenets of cleaning' that I have studied and propose as "best practice".

When a solvent is necessary to clean a fiber optic connection is when the contamination is a 'dry", or, when the contamination is a 'fluid' that requires emulsification (breakdown): a synthetic or mineral oil. I urge a solvent be used when inspection is not possible: you are cleaning in the 4th Blind-Cleaning Process.

WHAT MAKE AN IDEAL FIBER OPTIC CLEANER?

"Flammable or non-flammable". "Shippable". "Container or delivery type: pump, aerosol, pen, wiper". "Safe to use". "Environmentally safe". "Performs on a limited range of contamination types". "Performs on a wide range of contamination types".

Let me save you a little time: there is no one product that matches all these criteria, each of which is highly desirable. This means your selection is one of tradeoff. What you may find is the chemical that ships easily performs only on a limited range of contaminants, flammable works best, or, an aerosol is most effective. It is with this reality that the concept of 'applications specific cleaning' is further driven to the top of the list!

Ultimately, the only way you will know if any of these products actually work in your application is to try them. Your distributor may not be willing to send you a sample: I do not know of a manufacturer who would not be delighted to sample you their product. Why would they do this? It's simple: the manufacturer wants your feedback. The manufacturer wants to know if their product worked for your application. Most manufacturers create products in an applications specific way!

You may find that the product you least imagined you would purchase, is the best for your specific application, costs less, and is easiest to use once the manufacturer trained your crews! Also, increase your awareness of personal safety when using any chemical. Be aware, that some chemicals have been banned for sale or on 'phase-out' lists. Others, such as IPA, may be subject to local environmental regulations, which while possibly not enforced, can be a problem if a site visit determines you are not in compliance. The point is: be aware and don't buy based on marketing and sales promotions!

Be aware there are new chemicals coming to market. So-called "Aqueous Cleaning" is used for electric motors...and there are successful adaptions of this chemistry for fiber optics. Always ask: *"Are you aware of any new chemicals that would work for this application?"*.

The current range were formulated in the late 1980's and early 1990's. These chemicals were created in response to concerns about CFC's and Stratospheric Ozone Depletion. Did you know that IPA is a VOC and subject to regulations as an Atmospheric Ozone contributing substance. This brings what I consider one of the most important 'key thoughts" of this book: "Select a chemical that works best...using the smallest amount". Yes, this exists...and that selection depends on the contamination you are cleaning: this job will be different from the next job, which isn't the same as the one you did last week!

Please view the Appendix (Pages 55-56-57) with "Pros and Cons" for most cleaning products and devices. Call or email me if you have specific questions...I work in a 'vendor neutral' format and will be delighted to input.

Product Limitations: Swab Tools

Where would we be without the Q-Tip®? The cotton swab is a tool invented in the in 1923. After attaching pieces of cotton to tooth-picks, Leo Gerstenzang created the fist Q-Tip: "Q" stood for 'quality'. Nearly 100 years later, just about any swab may be called by the generic name: Q-Tip®...which must delight or confound the marketing people at global industrial power Unilever®!

Surely, in these times, there are many types of swabs: a.) "Clones of Q-tip" for cosmetics, b.) cleaning tools for dust in auto detailing, c.) swabs used in highly sophisticated environments: sterile medical procedures, critical production of solid state components, and yes...even precision cleaning fiber optics! All can be called a Q-Tip...but these precision tools are not made of wrapped cotton. In reality, even the very-best Q-Tip (there are several grades) would not be used in precision cleaning. _Why_? Cotton leaves a lint residue.

With the advent of 'probe tools' some felt that swabs for fiber optics would be obsoleted. Precision cleaning a V-Groove on a fusion splicer and the mirrors are cleaned with special swabs: a cotton tip for this applications, while recommended by some, leaves a residue. Surely cotton would not be used in a Class-1 Operational Cleanroom and never be used for fiber optic end face cleaning. So, what good is a fiber optic swab in these times?

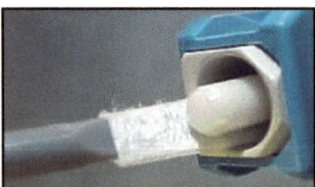

One process a 'probe tool' can't perform is cleaning the side of a ferrule as shown in this image. In fact, fiber optic swabs clean APC angle connectors as well...and maybe better than probe tools. _Why?_ Fiber optic swabs can have compliant tips conform to the bevel on the APC. Probe tools tips tend to be rigid.

When NTT introduced Cletop® in the late 1990's to clean 'jumper side', there was and is a companion product to clean the 'backplane side'. NTT-AT 2.5mm is a hand wrapped microfiber product. In 2003, Chemtronics® introduced a wrapped foam swab tool that had ridges to clean the alignment sleeve. Shortly thereafter a 1.25mm swab with a 'pull-truded' polyester was patented. MicroCare® followed with a 'sintered polyester' that avoided an IP standoff!

Now the questions arose: microfiber, polyester, foam...which works best? Each claims superiority. Microfibers are in ever-increasing use in clean room applications and despite protests about 'foam" that material is used in sterile medical as well as clean rooms. The battle between 'pull-truded' and 'sintered' polyester never gained marketing momentum as 'probe tools' convinced the end user that 'click and go' was good enough!

The issue with swab tools is...convenience...well, actually, inconvenience. Each swab should only be used one time. Grasping the small handles has been addressed by knurls and extension tips. There are numerous technicians who 'won't clean any other way. For others, the swab is always nearby when 'the other way doesn't work'! Recently I was asked to clean a military-style connection that the probe would not 'reach'. I was able to present three swabs to the end user as a 1st cleaning choice.

Obsolete? No, swab tools are another 'applications specific' fiber optic cleaning necessity that should be there as a back up. It's like having a 2nd screwdriver.

Sorry Unilever...a Q-Tip® always feels great to clean an ear or under-eye...but is not appropriate for fiber optic precision cleaning. If you are asking your distributor for a Q-Tip®, and cotton buds are in your took box...remember this can be why you are cleaning and cleaning and the system won't test! Q-Tip® is The Wrong Bud for this application!

Is First Time Cleaning Possible?

The existing standards permit 'up to five times' to clean a connection before replacing a jumper, re-termination, or, return of the circuit card to the OEM.

Let me be honest with you, I get a little 'flack' over suggesting "Blind Cleaning"! However, over more than 15 years of working in the field and most recently a formal survey conducted at various events over a 48 month period I have recorded that only 60% of you who are cleaning...are inspecting every time!

This is where the test-equipment folks make their sales-pitch to buy an inspection scope and I am one of those also! However, I also know you or your company may not have "budget", the battery died, or the one you have is on loan, or, there are 1,000 connections to PM and there isn't enough time!

It's for you...that I encourage the process change that leads to 'first time cleaning'. This is the 3rd procedure and published in Telcordia GR-2923. It's the newest thought from industry thought leaders.

Nothing is ever going to be 100% all the time! However, the December-2014 study I conducted, that compared cleaning ability of fiber optic cleaners, and many, many, thousands of real-time demonstrations at countless events leads me to state conclusively that a 'simple process change' in what you are doing now...makes your favorite cleaning tool work better.

Using a Cletop®...yes I can show you now to make it work better. Using a OneClick®, yes, I will show you how to make it work better. Using a cleaning platform from Stickers®, Chemtronics® or Seikoh Giken...yes, I will show you how to make it work better. Using swabs ... yes there is a place for what some call "obsolete".

Let's discuss all of these products and define how they are used: a process. If you had attended one of my sessions this next chapter would have included either a live demonstration or video.

I ask you now...please pretend we are face-to-face: go to YouTube:

https://www.youtube.com/channel/UC1a552-2i620UP6mM9WhwRg

1. **A DRY PROCESS ... that may only "move" debris and/or establish a static field that attracts additional debris**

2. **A 'WET-TO-DRY' PROCESS ... which may/may not actually dry the surface leaving unseen areas 'flooded' with excessive cleaning solvents**

3. **A "HYBRID OR COMBINATION PROCESS"** (noted in Telcordia GR-2923-Core, the most recent standard) **defines amount of fiber optic grade solvent and a drying technique.**

4. **A "BLIND-CLEANING© Process" leap-of-faith ... without benefit of adequate or any video inspection**

DRY PROCESS USING TOOLS, SWABS, AND WIPING MATERIALS

The pictures below are not 'one-off' or 'photo shopped'. One of the proofs of science is repeatability and this image has been captured countless times since it was first taken in 1999. You can emulate this by taking a dry wiper and cleaning the LCD in front of you...maybe cleaning mud or dust from your car.

The "dry cleaning process" is not effective for 'dry debris'. In fact, 'dry cleaning' is a mopping action used for 'wet contamination'. If you are using a fiber optic inspection device 100% of the time and can identify the contamination as a "fluid" then "dry cleaning" is an acceptable process choice. Please, check the video....

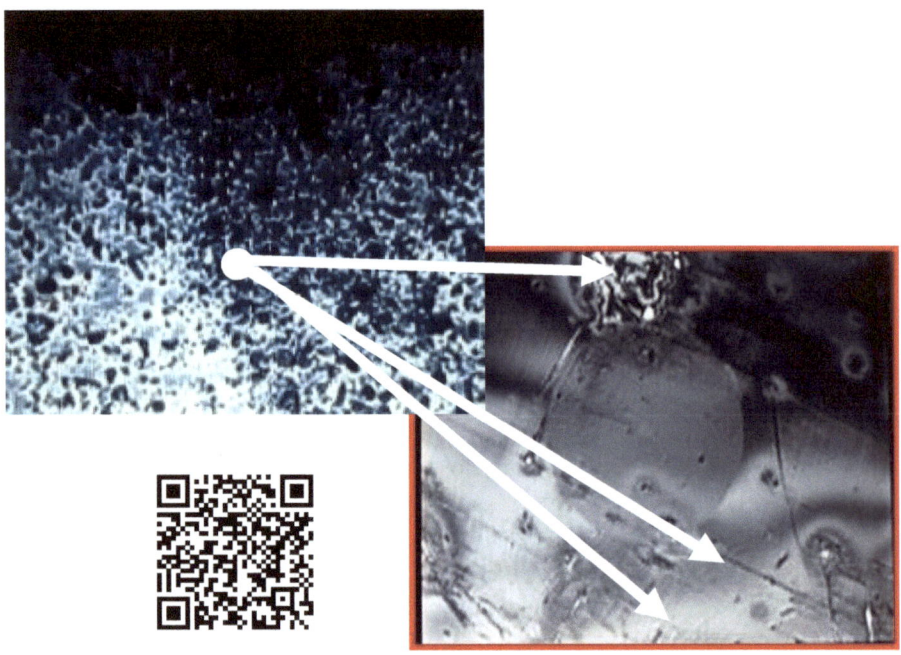

https://www.youtube.com/channel/UC1a552-2i620UP6mM9WhwRg

Dry cleaning can create an ESD tribo-charge which attracts additional debris. It does not require a large Charge: Chemtronics® tested using only 35-37 kv to contaminate to a significant level shown in this image;

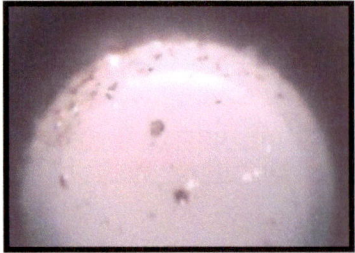

Dry cleaning tools are exceptionally convenient, easily fitting in shirt pocket or tool case. However, as fiber optics advances, so does the need to not just Clean conveniently but rather 'precision clean' to a higher standard. This new cleaning method takes no more time nor costs more than any other way.

It's a process change from old ways to new...following rapid advances in fiber optic technology of the last 20 tears...well into the future.

STATIC FIELD CONTAMINATION
FROM THE DRY CLEANING PROCESS

A quick check of "The Free Dictionary" reveals an 'urban idiom' we all have used: *"Don't give me any static."* However, there is 'good static' and 'no so good static'.

"Good Static" holds the screen protector to your tablet! In 1938, Chester Carlson invented the first xerographic image, pursing a concept of electrophotography. The result was XEROX Corporation with a heritage of printing…based on static attraction of 'toners' to a paper surface.

"Other static" is the type that damages sensitive electronics in much the same way as that door knob 'zapped you' the other day! Another type of static attracts dust to surfaces: perhaps this static field attraction is present on your computer screen!

That same phenomenon attracts dust to the ultra-small surface of a fiber optic end face.

Static, such as the attraction engineered into the screen protector and Mr. Carlson's amazing devices is engineered and predictable. As well, static that may damage electronics is also predictable: largely a result of temperature and humidity extremes. However, that is not the only reason "static happens".

EOS/ESD static is either conducted using grounding or dissipated using a solvent. In some instances, both techniques are used. ESD is a science unto its own. I urge you research by visiting the EOS/ESD Association: www.edsa.org if the topic fascinates you as it does many of us.

I consider static field contamination one of many potential types of contamination. As such, it is 'best practice' to include it as one that must be removed. However, static field contamination is not actually 'removed".

Static field contamination is prevented.

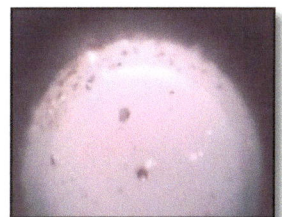

The dusty end face to the left was contaminated by drawing an end face over a dry wiper. The resultant static charge of only 31kv caused sufficient static field attraction to 'ugly up' this end face!

- **AS RELATIVE HUMIDITY DECREASES, STATIC FIELD ATTRACTION INCREASES**
- On a production line, relative humidity is controlled. ESD 'conducted' with devices.
- Relative Humidity 'outside' is infinite! ESD can't be managed with a wrist or heel strap: there is no conductive path
- Introduction of a fiber optic cleaning fluid creates a medium for the static field to 'dissipate'.

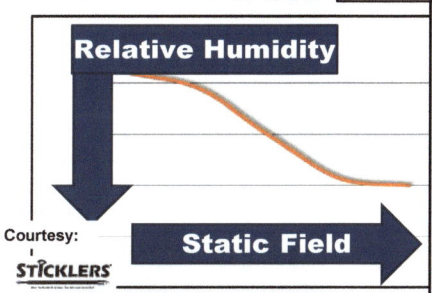

Courtesy:

STICKLERS

This test series from 2006 demonstrates that lightly moistening the same wiper resulted in significantly lower static field on the end face which did not attract dust.

Lab study Courtesy: **Chemtronics**

WHAT'S IN A WORD?

"Wet-to-Dry-Cleaning"

Everything in Moderation!

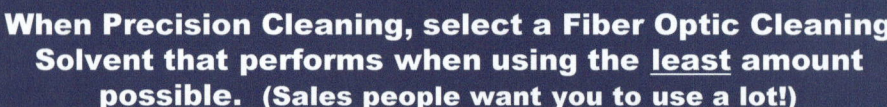

When Precision Cleaning, select a Fiber Optic Cleaning Solvent that performs when using the <u>least</u> amount possible. (Sales people want you to use a lot!)

Start cleaning anything...using small quantities.

"Wet-to-Dry" Cleaning

"What's in a Word?" What we say should be what we mean!
"WHERE'S THE BEEF?"

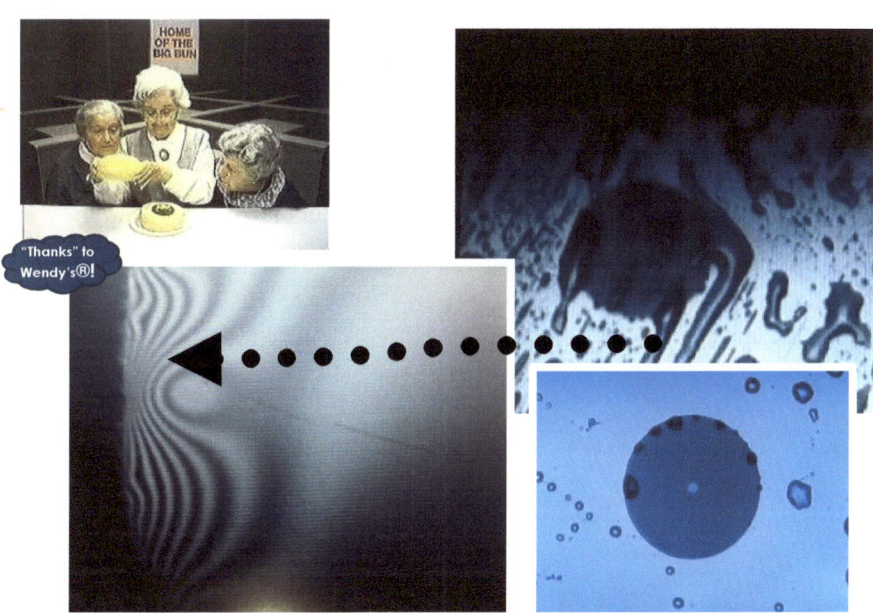

"Thanks" to Wendy's®!

The 'wet-to-dry' process may, or may not, actually dry the end face. It is with 'wet-to-dry' that the _inextricable interaction between fiber optic precision cleaning and precision inspection_ is most clearly defined. The images to the right are fluids on the end face. The image to the left is fluid that is being fed to the end face from outside the 'field-of-view'. The image to the left is captured using a MicroEnterprises ME-5000 video scope with a 'scroller' that enables a complete view of the end face.

The image bottom left shows the compete connector viewed with RMS-1 inspection. Take the time to see this video...it has been repeated hundreds, if not thousands of times. The video is an amazing revelation.

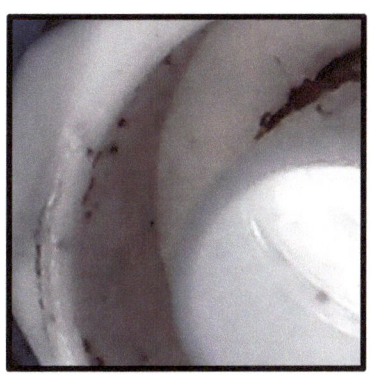

https://www.youtube.com/channel/UC1a552-2i620UP6mM9WhwRg

45

'Best Practice' Techniques

The 3rd Technique uses a high-performing fiber optic cleaner and non-woven wiper. The quantity of solvent is defined: less than 1ml (spot about the size of a USA quarter dollar) and the drying step moves the end face away from the initial point of contact.

The process assures that the end face is not "flooded" with excessive cleaner that can be trapped in the recesses of the connector...including Zone-5

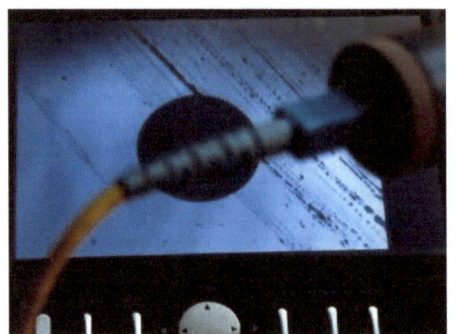

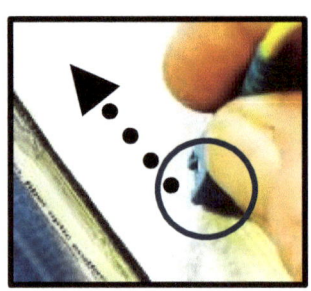

The Precision cleaning sequence starts with inspection of the end face. In this way you are aware of what you are cleaning, and the result. It seems obvious, but for many this is starting over.

A small amount of fiber optic grade cleaned and then the end face is moved from the wet-to-the dry. This is my original intent of 'wet-to-dry' and properly performed it is "Best Practice".

Vi https://www.youtube.com/channel/UC1a552-2i620UP6mM9WhwRg

The fiber optic "precision cleaning motion" is a _straight line action_ that moves debris and contamination _away_ from the initial point of contact. Another key benefit of the larger 'cleaning platform' is the surface area enables 'finding the angle' on APC connectors easier.

Proper cleaning means that any tool, probe, swab or platform meet the end face to the surface in a perpendicular to assure the compete surface is properly cleaned.

'Best Practice' Techniques

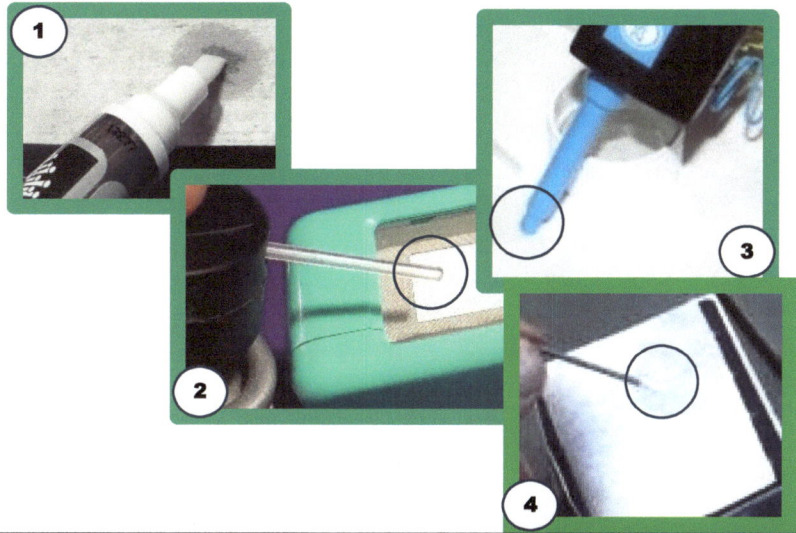

"HOW TO PROPERLY MOISTEN ANY FIBER OPTIC CLEANING TOOL"
Only use a small amount of a fiber optic cleaner. No IPA!

1.) Moisten the cleaning platform surface…about the size of a coin!

2.) Lightly moisten the 'reel cleaner' tape

3.) Depress the probe tool to moisten the cleaning tip. Don't activate … if it "cliks" the solvent won't be in the right place!!

4.) Hold the Probe or Swab tip in the moist area for a count of 1-2-3-4-5.

Don't:

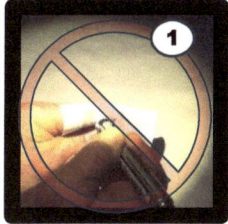

1.) _DON'T place a wiper over your finger or in the palm of your hand_. *Hand oils can transfer through the wiper to the end face.*

2.) _DON'T clean on a hard surface_. *A work table surface can 'grind-grit' and damage the end face.*

3.) _DON'T "dunk"!_ Contemporary fiber optic grade cleaners are formulated to perform using MINIMUM QUANTITIES. *It is possible to use too much!*

The "Figure-8" Motion is a "Polishing Action"

Fusion Splice Prep is not End Face Cleaning and yes, this is obvious, remember the two are 'applications-specific' operations. Train yourself and others about the differences.

The "Figure-8" is a polishing action. Take a look at what happens during this process compared to the 'straight-line' cleaning motion.

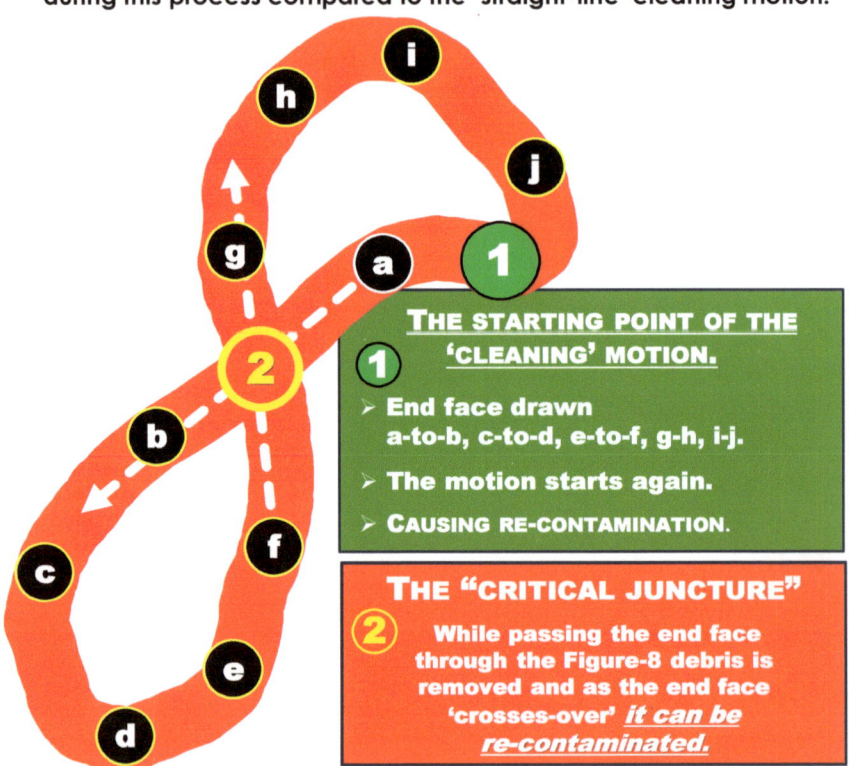

THE STARTING POINT OF THE 'CLEANING' MOTION.

1

➢ End face drawn
a-to-b, c-to-d, e-to-f, g-h, i-j.

➢ The motion starts again.

➢ CAUSING RE-CONTAMINATION.

THE "CRITICAL JUNCTURE"

2 While passing the end face through the Figure-8 debris is removed and as the end face 'crosses-over' *it can be re-contaminated.*

Product selection, process selection and actual techniques all play a part in the path to "Best Practice".

Since the first standards were issued in 1998, few have really studied what it actually means to precision clean and inspect. As fiber optic science is challenged by increased speeds over copper and co-ax, we all have a responsibility to assure success of not only each installation, but also the Industry as a whole,

Properly cleaning a fiber end face didn't seem to matter at a kilobit or megabit, but FTTp deployments and video over fiber proved proper cleaning and inspection is critical...and will be more so as we move though Gb/s through Tb/s...to ever higher capacities and consumer demands.

BEST PRACTICE IS <u>NOT</u> TO "DRY CLEAN FIRST" AND THEN FOLLOW WITH A "WET-TO-DRY"…IF THAT DOESN'T WORK!

Best Practice is to use a small amount of fiber optic grade precision cleaner with all cleaning tools and devices…the 1st Time!

How to Clean can be confusing!
Eliminate the Confusion-of-Choice!

What Works!

This group of products always used with precision fiber optic solvents is "Best-Practice"

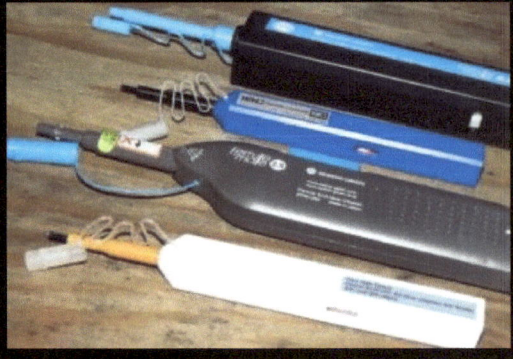

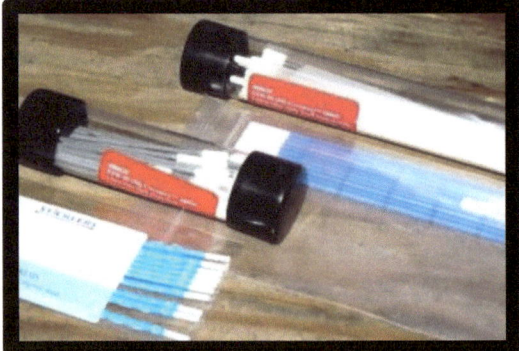

Be Prepared® ...those Brontobits are Coming! Start re-training nearly 20 years of missed-impressions...now!

UNTIL STANDARDS 'CATCH UP' :

✓ Designers help installers to 'self standardize'.

✓ Trainers ... it's time to update to 'Best Practice'

"Best Practice" ... has two components!

Inspection is critical.

However, if not available ... a procedure that cleans 'the first time' is a safety net <u>indication of effectiveness of the technique</u>.

Make 1st Time Cleaning your new "Best Practice Standard".

CHANGE THE PROCESS by always using a small amount of **non-IPA** fiber optic grade precision cleaner.

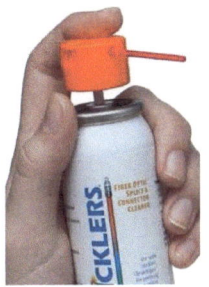
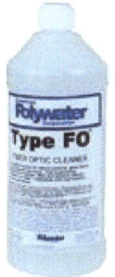

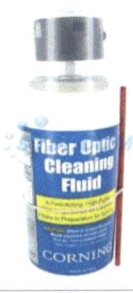

When you make a purchase of a fiber optic cleaner or inspection device ask: "Why"?

Challenge the supplier.
(Remember, your distributor may only be a sales agency ... request factory support if in doubt.)

Eliminate the

"Myths of Cleaning"

"How you clean and how you inspect" should never be "the problem".

The "sciences" are straight-forward.

The future of fiber optics depends on a surface and surrounding areas that are pristine clean.

ABOUT THE AUTHOR:

Thank you for purchasing this book. Ed Forrest has been actively involved in specification and applications engineering of various precision cleaning applications for more than 25 years. Previously employed at ITW Chemtronics®, he retired in 2014. He is schooled to analyze precision and gross cleaning applications in a wide range of applications. In 2001 he began development of a program that resulted in formal approvals at all major telecommunications providers.

He has seven patents specifically in the areas or fiber optic precision cleaning with six products in production. He also innovated, and patented, a chemical mid-span break-in for ribbon fiber. Those patents and others are assigned to Illinois Tool Works.

In Spring-2016 he began work on a means to prove this thesis of the three-dimensional nature of a connector and the impact of three dimensional contamination. While this is obvious, all standards are based on not 3-D, but only 2-D. Mis-understanding of cleaning and inspection is based on a 'flatland' approach.

Ed is active on fiber optic standards committees and is considered a SME in the study of fiber optic cleaning and inspection. His work is based on field experiences and the needs of designers, crafts persons and production line workers.

His practical thesis of "Five Zone Cleaning" is a look forward to the times when high speed and capacity of fiber optic transmission (even more) will be impacted by a contaminated or improperly cleaned connections. He has uniquely researched inspection of the 4th and 5th Zone and the influences of various debris and contamination positioned on areas of the connector. His new inspection device with a patent-pending rotating adapter© that enables digital photography of all surfaces has been heralded as innovative and past-due.

As an Electronics Manufacturer's Representative throughout the 1970's, he actively participated in the early introduction of some of the most fundamental electronic products in the changeover from analogue to solid state. These included solid state components, consumer products including hand-held calculators, esoteric high fidelity, test equipment, games and other electronic products (including the 'delightful insanity of CB-radios) considered 'cornerstones' of the contemporary marketplace. He also has production credits in that Industry

Turning to a then-developing market segment in the Home Furnishings Industry, by coordinating North American and International Development, using an effective agency in Denmark he was able to work throughout Europe prior to the time of the EU. In coordination with C.ITOH (est-1860) , he traveled and developed a Japanese market long before current interest in the important nations of The Pacific Rim. He initiated promotional activity in conjunction with USA Embassies, individual USA states resulting in active trade in Denmark, Sweden, Finland, Italy, Germany. Great Britain, nations in The Middle East and South Africa. Also, he has production credits in that industry segment.

Early career experiences as a Technical Representative in Union Carbide Corporation's Automotive Consumer Products group with career-forming experiences include: introduction of Prestone® AntiFreeze as a Summer Coolant in a one year NASCAR race test and associated promotions, as well as, an innovative time with Standard Oil of Ohio® as SOHIO® introduced "self-service fueling" to the market. He competed in this massive market when brands like STP® , BarsLeak®, Simonize® and Wynn's® dominated consumer interest.

He is a hobbyist collector of esoteric high-fidelity, enjoys photography, archaeology, ancient history and raced in SCCA 'wheel to wheel' in more than 300 events in cars of his design ... with a little help from his friends! . Married with a fascination for Weimaraners, he and his wife, Lanet, are often on the edge with three lovely specimens. They travel as they can. The picture is with "Bedford" who had retrieved a white brick!

QUESTIONS & CHALLENGES?

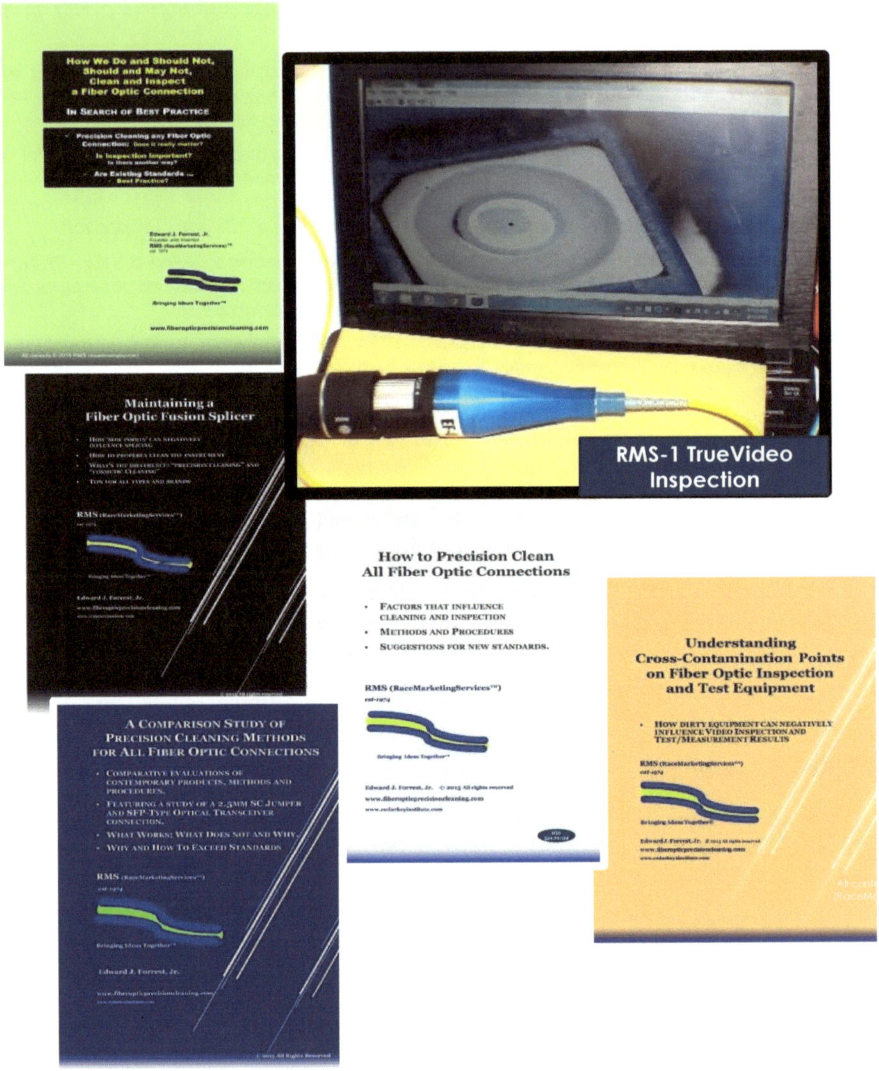

Ed Forrest:
Founder, Inventor, Author
RMS (RaceMarketingServices)™ est: 1974

----

edforrest@fiberopticprecisioncleaning.com
edforrest@live.com +770-971-8100

www.fiberopticprecisioncleaning.com

References:

1. Residual Contamination on Fiber Optic End Faces from Use of Polar Alcohols Paul Blair and Edward Forrest ITW Chemtronics June 2002 (rev:11/03) White Paper

2. HFE-7100/Proprietary Formulation Cleaning Comparison to 99.9% Isopropanol James Fitzgerald. ITW Chemtronics   Research Chemist: 2003. Laboratory Test Soils a.) Animal Fat – Representative of skin and fingerprint oils Multipurpose Grease b.) LUBRIMATIC #11316 Motor Oil c.) Quaker State 10W 40 d.) Silicone Oil Dow Corning representative of pulling lube and buffer gel

3. HFE-7100/Proprietary Formulation Cleaning Comparison to Precision Hydrocarbon/Proprietary Formulation James Fitzgerald. ITW Chemtronics Research Chemist: 2003. Laboratory Test Soils a.) Animal Fat – Representative of skin and fingerprint oils b.)   Multipurpose Grease – LUBRIMATIC #113163) c.) Motor Oil Quaker State 10W 40 d.) Silicone Oil – Dow Corning representative of pulling lube and buffer gel

4. A Study of 99.9% Isopropanol Absorption Rate of Water from Air Susan Max: Lead Chemist. ITW Chemtronics® 2004. Laboratory Test.

5. A More Effective Means of Cleaning Fiber Optic Connections in FTTH, Outside Plant, and, OEM Applications. White Paper-Edward J. Forrest FTTH-2005

6. Inspection and Cleaning Procedures for Fiber Optic Connections All contents © 1992–2006 Cisco Systems, Inc. Document ID: 51834 8-26-2006

7. Soil Removal from End Face Utilizing Cisco Series of Ten Diverse Soils: Paul Blair, Ed Forrest, And Susan Max ITW Chemtronics: 2006. Laboratory Test

8. Generating a Static Field When Precision Cleaning a Fiber Optic Connection: Paul Blair, Susan Max, and Edward Forrest. ITW Chemtronics: 2008 Lab Test

9. TIA 455-240 September-2009

10. IEC 61300-3-35 ed1.0 2009 August 2009

11. Telcordia GR-2923-CORE. February-2010

12. Contemporary Considerations When Precision   Cleaning Fiber Optic Connections: Performance-Inspection-Environmental Matters: White Paper Edward J. Forrest: November-2010

13. SAE AIR 6031. 2012 Cleaning fiber optic connections.

14. Comparisons of various cleaning solvents acting on ten complex soils and Investigation of Contamination of the Horizontal and Vertical Ferrule. Laboratory Test recorded on Video. Edward J. Forrest: January 2011

15. Interferometer readings courtesy of Promet Corporation using a FiBO® 250 device.

16. Contemporary Considerations When Precision Cleaning a Fiber Optic Connection: 2011v8

17. Edward J. Forrest: Power Point. "Comparisons of cleaning techniques with audio and video". Training session. 2014

18. Bill Woodward "FOI" Fiber Optic Installer (ETA) and published by SYBEX

19. "Video Lab Comparisons of Aqueous Cleaning Products and Procedures": Edward J. Forrest, Jr. August-2015

20. "VideoLab Tests of Various Cleaning Procedures on simple and complex contaminants." Edward J. Forrest, Jr. August-2015

21. "Clean and inspect: What IEC 61300-3-35 means to you", *Cabling Installation and Maintenace*-July-2015. Brian Teague, MIcroCare.

22. "The Need for Processes that Future Proof the Fiber Optic Installation". White Paper. Edward J. Forrest, Jr. July-2015

23. "An Evaluation of Aqueous Cleaning Processes for Fiber Optic End Face Connections". VideoLab Study of commercial and experimental aqueous cleaners. PowerPoint Edward J. Forrest, Jr. August-2015

24. "New Inspection Criteria for all Fiber Optic Connections". White Paper. Edward J. Forrest, Jr. Summer-2016

25. "The many problems of cleaning with IPA". *Cabling and Installation Maintenance*-October-2016 Edward J. Forrest, Jr.

26. "Using Precision Cleaning and Inspection Processes to Future-Proof Optical Fiber Installations-A Search for Best Practice. *ICT Today*. Edward J. Forrest, Jr. Fall-2016

27. "How we do and should not; should and may not, precision clean and inspect a fiber optic connection". Training with video. Edward J. Forrest, Jr. 10/2016

PRODUCT STRENGTHS AND LIMITATIONS
THE CLEANING TOOLS

Product	Strengths and Limitations
Cleaning Platforms	<ul><li>Only clean jumper side or accessible connections from the rear of some equipment</li><li>Larger cleaning surfaces moves contamination away from initial point of contact. Very clean process.</li><li>Larger containers tend to be inconvenient. Research smaller sizes.</li><li>Disposable components, lowest costs per cleaning.</li></ul>
Probe Tools Sticks and Swab Tools.	<ul><li>Probes: Exceptional convenience cleans both jumper side and back plane</li><ul><li>Suggest moisturizing before use</li></ul><li>Smaller cleaning surfaces may not absorb effectively</li><ul><li>Some devices have larger surfaces than others</li></ul><li>May not clean alignment sleeves and other connector "geometry"</li><ul><li>Side wall of cleaning probe cleans alignment sleeve with same wiper surface as the end face.</li></ul><li>May require multiple cleaning passes</li><ul><li>Strongly suggest multiple passes</li></ul><li>Swabs...no cotton! Cleanroom Microfiber, foam and special pull-truded or sintered polyester fine as are non-woven cellulose/polyester</li></ul>
Cassette, Reel Cleaning Tools	<ul><li>Exceptional convenience, refillable, cost effective</li><li>Smaller cleaning surfaces not as effective on wide range of contamination</li><li>May require multiple cleaning passes</li></ul>
Mechanical Cleaning Devices	<ul><li>Larger systems more suitable for OEM production</li><li>Higher cost of acquisition</li><li>Inadequate solvent may leave residues that surface bond to the end face making contaminates more difficult to remove.</li></ul>
ALWAYS CHALLENGE THE SUPPLER... **COMPARE AND CONTRAST**	Obtain a sample...but do "them" a favor: _tell them how it worked for you!_

Courtesy of: STICKLERS

Chemtronics®

PRODUCT STRENGTHS AND LIMITATIONS
FIBER OPTIC CLEANING SOLVENTS AND WIPERS

Product	Strengths and Limitations
1. **Isopropyl Alcohol**	1. Hygroscopic, nature further diminishes limited cleaning ability. Flammable. Aroma. Has shipping restraints.
2. **Precision Hydrocarbons**	2. Highly Flammable, exceptional cleaning ability, aroma. May have some environmental limitations. Has shipping restraints for some containers.
3. **HFE-7100**	3. Non-flammable, limited cleaning ability compared to hydrocarbons, increased environmental scrutiny. Aerosols may have shipping restraints.
4. **"Aqueous Cleaners"**	4. Non-flammable. Must be actively dried. Excellent cleaning ability. EZ Shipping.
Pumps and Squeeze Bottles. Pens. Aerosols.	1. Squeeze containers and pumps draw in air: may cause cross contamination or degradation of the cleaning solvent 2. Pens are one-way. Make sure tips are clean 3. Aerosols containers are the cleanest one-way delivery system. Shipping per federal and international statues commonplace, but strictly regulated.
Compressed Gas Dusters	1. Not appropriate for precision cleaning a fiber optic end face. 1. Will not remove oily contamination 2. Consider as a 'pre-cleaning' step for water damaged connections 3. Does not create a static charge
1. **Natural Fiber Products (Paper, cotton)**	1. Linting caused by shredding paper or cotton fibers
2. **Non-Woven Materials**	2. Stronger than paper but not all are as absorbent. Preferred over 100% or cotton products.
3. **Microfibers**	3. Always select cleanroom grade microfiber materials for fiber optic applications

Seek 'trade-offs' that result in
Best Practice for each installation.

Courtesy of:

PRODUCT LIMITATIONS:
OTHER CONTAINERS AND WIPERS

Product	Limitations
Pumps and Squeeze Bottles	<ul><li>Uncontrolled environment</li><li>Pump or Squeeze draws in air causing cross contamination and product dilution</li><li>Discard remaining IPA each day to hygroscopic contamination</li></ul>
Moistened Lens Grade Tissues	<ul><li>Wipe materials may not be optical grade</li><li>Solvents may not be optical grade</li><li>Do not use wipers intended for microscopes, eye glasses, camera lenses on fiber optic surfaces.<ul><li>Check carefully</li></ul></li></ul>
Natural Fiber Products (Paper, cotton)	<ul><li>Lost fiber residues</li><li>Linting and shredding</li><li>Varied absorbency</li><li>Limited tensile strength for paper wipers</li><li>"No such thing as lint-free paper or cotton"!</li></ul>
Polyester, Nylon, and, Microfibers, non-woven hydroentangled polyester-cellulose blends.	<ul><li>100% polyester and nylon is very strong but not absorbent<ul><li>Many types: the best practice is to avoid this group as there are better selections in non-woven and some microfiber groups</li></ul></li><li>Only select 'cleanroom-grade' microfiber materials. These have been vetted by industries with critical cleaning requirements.</li><li>Select non-woven hydroentangled materials carefully. Major producers supply this wiper in some reel cleaners and all cleaning platforms.</li></ul>
Air Compressors	<ul><li>Typical for production line operations</li><li>Although heavily filtered...may depart residues from the equipment.</li><li>Not ideal for FTTx operations</li></ul>

Courtesy of: STICKLERS

Chemtronics®

TEST YOUR KNOWLEDGE:

1.) ___T___F   The first thing a technician should do is observe the ambient environment and get an idea of the type of debris or contamination may be present.

2.) ___T___F: 99.9% isopropyl alcohol (IPA) is the best cleaner because it removes the widest range of debris and contamination all the time.

3.) ___T___F: Cotton and paper are just as good as microfibers and non-woven cellulose/polyester blends to hold and absorb debris and contamination.

4.)___T___F: "Field of View" refers to the area seen by an inspection device.

5.)___T___F: Fluidic contamination stays in place and dry debris moves. There is not a transfer between end faces.

6.)___T___F: One problem with swabs is that foam is not as good as other materials.

7.)___T___F: All fiber optic cleaning solvents are the same; simply packaged differently

8.)___T___F: If I do not have a video scope, cleaning multiple times works just fine

9.)___T___T: A major advantage of "probe tools" is their convenience

10.)___T___F: When using a cleaning platform, hold the end face at 90 degree perpendicular to the cleaning surface for best result.

11.) ___T___F: The "best practice" fiber optic cleaning technique is a "Figure-8" motion because it is fast and repetitive.

12.) ___T___F: "Wet to Dry" cleaning does not require video inspection.

13.) ___T___F: The straight line cleaning action moves dry debris and fluidic contamination away from the initial point of contact.

14.) ___T___F: Fiber optic standards, such as IEC 61300-3-35, are updated every 5 years with annual bulletins and updates.

15.) ___T___F: Many of the same cleaning techniques and products typically used for fusion splice prep are appropriate for end face cleaning.

16.) ___T___F: Class-1 Direct View Microscopes are acceptable to observe a live fiber

17.) ___T___F: A db loss test set can indicate if the end face is clean.

18.) ___T___F: Network designers should not be involved in deciding how to clean and inspect.

19.) ___T___F: Anyone can participate in the international standards process

20.) ___T___F: Expanded beam connections do not require cleaning

21.) ___T___F: Static field attraction of dust is possible using the dry cleaning process

Additional Learning Answers

1.) <u>True</u>: *Knowing what is being cleaned is obvious, but often ignored as 'blind cleaning' without inspection or inadequate inspection is more common than we like to admit!*

2.) <u>False</u>: 99.9% isopropyl alcohol is not as good a cleaner (solvent) than a precision hydrocarbon, HFE7100 formulas, and new aqueous cleaners.

3.) <u>False</u>: Cotton and paper lint, tear and shred. These are good enough to absorb a fluid...but not for precision cleaning. In doubt...don't use them.

4.) <u>True</u>: "Field of View" is relative and limited by magnification. 400x actually sees less of the end face than 200x. Resolution also is an important characteristic of a video scope. Higher the magnification, greater the resolution. Look for lower magnification and greater resolution for "best practice". The more you see, the more reliable.

5.) <u>True</u>: Fluidic contamination moves and transfers while dry debris stays in place. Trick question because debris transfers!

6.) <u>False</u>: There are literally thousands of types of foam. Be sure you select ones that are "medical" or "cleanroom grade". Foam absorbs and can be a great cleaning media.

7.) <u>False</u>: There are even numerous HFE7100/7200 formulas and all of them are not the same cleaning power. There are also HFC, Precision hydrocarbons and aqueous cleaners. Some cost about the same as 99.9% IPA, others more expensive. Select in applications-specific way...not all work the same, even from the same producer.

8.) <u>False</u>: If a video scope is not available...always use the 3rd "hybrid/combination" process as the safety net. Cleaning can be a first-time event.

9.) <u>True</u>: Convenience is nice...but make sure the way you use the tool is effective.

10.) <u>True</u>: "Finding the angle" on a cleaning platform is done by observing a 'no snag glide' which is apparent when the end face meets the cleaning surface at a 90^0 perpendicular.

11.) <u>False</u>: The "Figure-8 Motion" is a polishing action for connector re-termination. It is not an end face cleaning action.

12.) <u>False</u>: Every connection, no matter which technique, should be video inspected. However, the "hybrid" or "combination" technique works best for all cleaning devices.

13.) <u>True</u>: Don't twist and turn...move the end face away from the initial point of contact.

14.) <u>False</u>: As of 2016, fiber optic standards, such as IEC 61300-3-35 are updated about every ten years.

15.) <u>False</u>: Fusion splice prep is a completely different application than end face cleaning. For fusion splice, the side and length of the fiber is cleaned, not the end face! Cross-use of IPA to end face cleaning can result in inferior results.

16: <u>False</u>: Never take a chance with your eyesight. Use 'direct view' to QC jumpers in the warehouse or on a mass production line...never in actual use.

17.) <u>False</u>: Only video inspection gives a clear decision of what is actually clean.

18.) <u>False</u>: The network designer is the 'first person' who understands the limitations of the deployment and should also specify an inspection and cleaning process to the installer.

19.) <u>True</u>: While it is not an easy task to become involved, most standards bodies are open to anyone. Some have a fee, others do not.

20.) <u>False</u>: Any contaminant on any surface of any connector can move and cause signal degradation or loss. Cleaning the entire surface, especially the 'fiber core' and other areas that can contaminate it in the time of post cleaning and post inspection is 'best practice'.

21.) <u>True</u>: Static contamination is one of many types of debris that must be considered when you specify a cleaning process or perform it.

WITH APPRECIATION:

How does inspiration arise?   Sometimes it's in the middle of the night and once in the shower!   This time it is clear: I received a call to photograph expanded beam connectors for an upcoming standard.   Bill Woodward, a mentor and author of a supreme book, *"Fiber Optic Installer"* encouraged development of RMS-1. We met 'half-way' and he ignored the 'rough nature of the original prototype' and encouraged the product that resulted in most of the images in this book. "Thank you, Bill."

No work of this detail could be accomplished without many inputs. These surely came from individuals such as Eric Martini and Paul Blair at ITW-Chemtronics®. Eric's understanding of 'solvent cleaning'; Paul's knowledge of 'wiping materials' provided critical inputs during my time there. Both led me to create.

I want acknowledge the many, many thousands of technicians who I met at trade shows and over hundreds of dozens of donuts at 06:30 in all parts of the world: as *well, without your input, I would not have learned.* I thank those who supported my work with formal approvals: PIDS, CIFA, PeopleSoft, and SSI numbers all drove my research to ever higher levels. I want to thank Dr. Osman Gebizlioglu and Dr. Tatiana Berdinskikh for their patience and challenges that helped me evolve through these thesis and continued study.

Another early influence is Glenn Porter of Microenterprises®: he remains so to this day. For many years I studied with his ME-9600. I use a modified ME-5000 to check my work. Oddly, both models are out of production. Porter's very early study of fiber optic microscopy and efforts to create an automatic, computer-assisted inspection system, long before the current generation of devices, deeply influenced the thesis of Five Zone Inspection, matters such as the three-dimensional aspects of the complete connector and debris. Having been trained to clean everything from a cooling system to medical pacemaker, transitioning to dry debris, fluidic contamination, and, combinations of the types for fiber optic applications remains in my life-long-career-DNA!

Mike Schneider, founder of Optical Design Manufacturing (ODM) carried forward some of Porter's work with his own innovative design: a range of portable instruments with wide "field of view" and astoundingly brilliant resolution that (at times) seems to show images like an interferometer. Schneider and Porter never met, to the best of my knowledge. However, their arrival at similar results, from different perspectives, is a critical tenet of scientific proof. I thank Sean Adam of AFL for his input.

There are others: Jim Hayes at The Fiber Optic Association and Larry Johnson, Founder of The Light Brigade have been not only encouraging, but also, highly influential for me...even when they may have disagreed!   Larry and Jim have contributed immensely to our Industry. Through the years I have been privileged to exchange ideas with Richard Ednay of OTT in the United Kingdom. As well, Frank Giotto of Fiber Instrument Sales and his team that includes Kim Teesdale, Kirk Donley and John Bruno have impacted the market and influenced my thoughts in multiple ways.

John Cotterill at JSC Aeroptics, also in the United Kingdom, always offered the best advice with an amazing sense of humour! *Others on the SAE AS-3 Committee include Larry Wesson, Don Stone, good people at BOEING, ARINC, NAVAIR, iNEMI, and IEEE-Aerospace all challenged me to higher levels.*

We all owe a debt of gratitude to Industry Training Associations: BICSI, ETA, FOA, and SCTE play a critical and essential role to assure technicians, designers and trainers are updated, certified and properly trained. I am appreciative to these groups and independent training providers who have provided me a forum to express my thesis and concepts. Trade publications such as *Cable Installation and Maintenance, ISE (formerly OSP), Fibre Systems, ICT-Today* provide invaluable updates in the time-between standards are updated and developed. As well, there are others-of-influence who cannot be named for reasons those in the Industry understand well.

In recent times I have been honored to associate with former "competitors and foes"!   Among these are Brian Teague at MicroCare who has provided important insights.

Finally, in the first place, I want to thank my wife, Lanet, and others in the family who endured the times during all hours of the night, perhaps on vacation, I would be tapping keys on the computer as they calmly closed the bedroom door in places all over this wonderful world!

THANK YOU, ALL.

December- 2016

After thoughts:

When the first cleaning tools were introduced to the market in 1998-99, it's clear that few could have foreseen the impact of fiber optic deployments. Fiber to the home didn't exist and CATV had not deployed DOSCIS© as a counter measure to Verizon's vision with FiOS®. Wireless streaming didn't exist. Concepts of 'back haul' was still in the future.

There are countless examples of how fiber optics, in a very positive and wonderful way, have influenced our lives and will for many years into the future. For example, the industry has not quite justified download that does not match upload ... there is along way to go.

I am troubled that for the last 20 years the sciences of cleaning and inspection have been 'under-trained'. Word-of-mouth and commercial positions have clouded important and essential tenets in our Industry as in no other. While my career is ending, I contribute and have training programs on license for reasonable fees. I will also conduct evaluations in a strict vendor-neutral manner should you wish to develop a curriculum or new product.

My new products, RMS-1 and RMS-2 are priced to enable more and better inspection by more technicians. It's impossible to say how many technicians, over the years shook their head *'no'* when I asked if they used video inspection. Many still believe that 'having a light' means the connection is clean. More so, a manager would ask me *'don't talk about video inspection...we don't have funding'*.

Thank you for reading this far...please...always feed back impressions...positive or not so much! That is what makes <u>us</u> better with the ability to future proof, change perceptions and eliminate the myths that protects the future of this wonderful communications medium.

All the best...just the best!

21 November, 2016

All contents copyright:
All Rights Reserved

Notes

www.ingramcontent.com/pod-product-compliance
Lightning Source LLC
Chambersburg PA
CBHW040848180526
45159CB00001B/355